Electromagnetic Radiation in Analysis and Design of Organic Materials

Electronic and Biotechnology Applications

Electromagnetic Radiation in Analysis and Design of Organic Materials
Electronic and Biotechnology Applications

Edited by
Dana Ortansa Dorohoi, Andreea Irina Barzic,
and Magdalena Aflori

CRC Press
Taylor & Francis Group
Boca Raton London New York

CRC Press is an imprint of the
Taylor & Francis Group, an **informa** business

CRC Press
Taylor & Francis Group
6000 Broken Sound Parkway NW, Suite 300
Boca Raton, FL 33487-2742

First issued in paperback 2019

© 2017 by Taylor & Francis Group, LLC
CRC Press is an imprint of Taylor & Francis Group, an Informa business

No claim to original U.S. Government works

ISBN-13: 978-1-4987-7580-9 (hbk)
ISBN-13: 978-0-367-88956-2 (pbk)

Contents

SECTION II Interactions of Electromagnetic Radiations with Macromolecular Materials

Editors

Dana Ortansa Dorohoi, born in Iasi, Romania, in 1945, is Professor Emeritus at the Alexandru Ioan Cuza University, Faculty of Physics. She obtained her PhD in physics in 1978 and gained the academic title of professor in the Department of Optics and Spectroscopy at the same university in 1998. Subsequently she worked on her habilitation and supervised so far around 25 PhD theses. In 2012, she was awarded with the title of Professor Emeritus of the Alexandru Ioan Cuza University. In parallel with her didactic activities, she conducted scientific research on various topics, including liquid crystals, simple liquids, optics of crystals and liquid crystals, intermolecular interactions in solutions, and electro-optical parameters of the molecules. The scientific contributions developed in 43 years (since 1972) by Professor Dorohoi included the interdisciplinary research for theoretical and experimental substantiations of the processes occurring in anisotropic media under electromagnetic radiations, thus developing new optical and spectral analysis methods of optical properties of materials in solution or solid state. She published well over 165 ISI-indexed articles in the fields of physics of liquids and anisotropic media, macromolecular science, photochemistry, and radiation chemistry and is the author of three patents, over 10 books or book chapters, and over 35 research projects.

She is a referee for several scientific journals in the area of optics and molecular spectroscopy.

Andreea Irina Barzic, born in 1982, in Iasi, Romania, has been a scientific researcher at the Department of Physical Chemistry of Polymers from Petru Poni Institute of Macromolecular Chemistry (Romania) since 2005. She received her PhD in chemistry from Romanian Academy in 2009 and in 2014 she finalized her PhD in Physics Emeritus at the Alexandru Ioan Cuza University, Faculty of Physics. Her major fields of interest include liquid crystals, optics of polymers, and structuring of materials under electromagnetic and shear fields. She has published over 35 papers in peer-reviewed journals and was an author of 12 books or book chapters. She is also a member in several research projects and referee for a number of prestigious journals in the field of physics and material science.

Magdalena Aflori, born in 1973, in Iasi, Romania, is a scientific researcher at the Department of Polymer Materials Physics from Petru Poni Institute of Macromolecular Chemistry (Romania). She graduated her PhD in physics in 2005 at Alexandru Ioan Cuza University, Faculty of Physics. She has remarkable scientific contributions in plasma physics, x-ray diffraction, small-angle x-ray scattering, polymer chemistry, and surface science. Her interdisciplinary research is quantified in more than 65 papers in peer-reviewed journals, seven books or book chapters, and one patent. She is a member in many research projects with national and international funding and referee for a number of international journals with a high-impact factor.

Introduction

Electromagnetic Radiation in Analysis and Design of Organic Materials: Electronic and Biotechnology Applications can be a reference work on theoretical, experimental, and practical issues involved in interaction of radiations with matter. The book attempts to cover several physical properties of materials in the presence of radiations from a wide range of electromagnetic spectrum, adopting a multidisciplinary approach to bridge physics of condensed matter, photochemistry, photophysics, and materials science.

The book explores a great variety of compounds highlighting the different approaches of the physical properties when radiations are interacting either with low molecular compounds or with polymers. To provide a systematic overview of how both types of materials respond differently to radiations, this book is divided into two sections.

The first section of this book discusses the interaction of electromagnetic radiations with low-molecular-weight materials, revealing that optical and spectral properties of these compounds in solution or solid state can be easily investigated through the data extracted after their interaction with ultraviolet-visible (UV-VIS) radiations. Chapter 1 is devoted to the presentation of the spectral methods that are used to estimate the dispersive forces in nonpolar liquids. The energy of dispersive forces in nonpolar liquids is determined by spectral means. Using the dependence between the spectral shifts measured relative to the gaseous phase of the spectrally active molecules, the polarizabilities in the excited electronic state molecule. This kind of evaluation is important for the development of quantum mechanical characterization of the excited states of the molecules. Chapter 2 contains the newest contributions to characterize intermolecular interactions in ternary solutions of some 1,2,4-triazolium ylids under optical radiations. Using the cell statistical model of ternary solutions, the energy in molecular pairs of the types ylid–protic solvent and ylid–aprotic solvent molecules is estimated. The difference between the energies in molecular pairs in the ground electronic state of interacting molecules is hardly estimated by other methods. The analysis through spectral methods of the intermolecular interactions in solutions of zwitterionic compounds is presented in Chapter 3. The solvatochromic features of some dipolar compounds in solutions prepared in solvents of various types allow one to evaluate some types of intermolecular interactions in solutions and to establish their contribution to the spectral shift measured in electronic absorption and fluorescence spectra. Some molecular descriptors in excited states of spectrally active molecules were estimated from solvatochromic analysis. Chapter 4 deals with the relevance of the molecular descriptors for the modeling/discrimination of amphetamines using artificial neural network. The efficiency with which each network identifies the class identity of an unknown sample was evaluated by calculating several figures of merit. The results of the comparative analysis are presented. The aspect of double refraction of light in transparent uniax anisotropic media using the channeled spectra is reviewed in Chapter 5. The contribution brings forward multiple ways to determine the birefringence of uniax anisotropic media from the study of channeled spectra. Different anisotropic layers of various thicknesses were presented and analyzed taking into consideration the birefringence dispersion. When passing from uniax anisotropic crystals to small-molecule liquid crystals, the birefringence must be differently approached. So, Chapter 6 contains new approaches on birefringence dependence on visible radiation wavelength for developing interferential optical filters. Thus, some methods for investigation in the case of liquid crystalline sample are described and some results in measurement of the dispersion birefringence of this type of anisotropic layers are illustrated. Applications in obtaining interferential filters are also presented. The book continues with the presentation of double refraction of light in biaxial crystals, as shown in Chapter 7. The methods used in optical characterization of the anisotropic inorganic crystalline layers are described and examples of applications are detailed. The method to determine the main refractive indices for anisotropic biaxial crystals is presented in this chapter.

The second section of this book is devoted to the interactions of electromagnetic radiations with macromolecular materials. Chapter 8 deals with the interaction of plasma radiation and particles with urinary catheters based on polymers in order to adapt their surface properties for antimicrobial purposes. To fulfill the biomedical requirements, polymers can be subjected to other types of radiations, as described in Chapter 9. A review of the most common effects of gamma irradiations on polymer materials that can be used in biomedical applications is made here. The effects of gamma irradiation on physicochemical properties of polymer materials are described. The importance of gamma irradiation is underlined and some applications are described. When polymers are designed for drug delivery purposes, one can utilize UV-VIS radiations in analysis of the release rate of the active substances, as presented in Chapter 10. In addition, Chapter 11 continues with a description of the interaction of these radiations when the polymer matrix is subjected to stretching and dichroism gives information on the preferred release direction of drugs like those used in the treatment of Alzheimer's disease. Thus, polymer foils of a hydrosoluble polymer with the drug donepezil included were obtained for two polymeric forms of donepezil (A and B). The influence of the foil-stretching degree on the mechanism of the drug release was emphasized by spectral means. Chapter 12 shows that the applications can be multiple when polymers are interacting with laser radiations. Surface structuring through such procedure can be viewed as a tool for micro- and nanotechnologies. Therefore, the chapter presents laser irradiation technique and examples of common lasers used in polymer science. Also, the fundamentals of interaction of the laser beam with matter are described. Development of the nanostructured surface relief created during laser irradiation process and its practical importance is presented. Chapter 13 contains aspects on liquid crystal polymers under mechanical and electromagnetic fields. The chapter starts with basic concepts, such as definition, classification, transition phases, and further progresses with the behavior of such materials under the influence of the above-mentioned factors. Starting from the presented physical characteristics, the applications in several industries are presented.

The topics presented in this book are addressed to those working as researchers, PhD students in doctoral or post doctoral school as well as engineers, and they can be considered as the source of information on all mentioned aspects. The physical properties performance of these materials or their design can be facilitated through the exposure of radiations. In this context, the aim of this book is to describe and interpret the interactions between electromagnetic radiations and matter for developing new methods of characterization or to adapt their shape and properties to the requirements of advanced technologies.

MATLAB® is a registered trademark of The MathWorks, Inc. For product information, please contact:

The MathWorks, Inc.
3 Apple Hill Drive
Natick, MA 01760-2098, USA
Tel: 508-647-7000
Fax: 508-647-7001
E-mail: info@mathworks.com
Web: www.mathworks.com

Contributors

Magdalena Aflori
Department of Polymeric Materials
    Physics
Petru Poni Institute of Macromolecular
    Chemistry
Iasi, Romania

Ecaterina-Aurica Angheluţă
Faculty of Physics
Alexandru Ioan Cuza University
Iasi, Romania

Mihai-Daniel Angheluţă
Iuliu Haţieganu University of Medicine
    and Pharmacy
Cluj-Napoca, Romania

Andreea Irina Barzic
Department of Physical Chemistry of
    Polymers
Petru Poni Institute of Macromolecular
    Chemistry
Iasi, Romania

Ana Cazacu
Department of Sciences
University of Agricultural Sciences
    and Veterinary Medicine
Iasi, Romania

Valentina Closca
Faculty of Physics
Alexandru Ioan Cuza University
Iasi, Romania

Dan Gheorghe Dimitriu
Faculty of Physics
Alexandru Ioan Cuza University
Iasi, Romania

Mihaela Dimitriu
Faculty of Physics
Alexandru Ioan Cuza University
Iasi, Romania

Dana Ortansa Dorohoi
Faculty of Physics
Alexandru Ioan Cuza University
Iasi, Romania

Irina Dumitrascu
Faculty of Physics
Alexandru Ioan Cuza University
Iasi, Romania

Leonas Dumitrascu
Faculty of Physics
Alexandru Ioan Cuza University
Iasi, Romania

Steluta Gosav
Chemistry, Physics, and Environment
    Department
Dunarea de Jos University
Galati, Romania

Nicolae Hurduc
Department of Natural and Synthetic
    Polymers
Gheorghe Asachi Technical University
Iasi, Romania

Cristina-Delia Nechifor
Department of Physics
Faculty of Machine Manufacturing and
    Industrial Management
Gheorghe Asachi Technical University
Iasi, Romania

Iuliana Stoica
Department of Polymeric Materials
    Physics
Petru Poni Institute of Macromolecular
    Chemistry
Iasi, Romania

Beatrice Carmen Zelinschi
Faculty of Physics
Alexandru Ioan Cuza University
Iasi, Romania

Section I

Interactions of Electromagnetic Radiations with Low Molecular Weight Materials

1 Spectral Methods for Estimating Molecular Polarizability in Nonpolar Solutions

Mihaela Dimitriu

CONTENTS

Abstract

The dispersion interactions are dominant in nonpolar solutions; they determine spectral shifts of the electronic absorption bands. The contribution of the dispersion forces can be separated from the total spectral shift recorded in a given solvent in the case of solute molecules with great differences between their polarizabilities in the electronic states participating to the electronic transitions. In this chapter, we discuss the procedure used for obtaining information about the dispersive interactions in some solutions in order to estimate the molecular polarizability in the electronic excited states. This kind of information is useful in quantum mechanical chemistry of the excited molecular states.

1.1 INTRODUCTION

In each moment, a molecule possesses an instantaneous electric dipole moment created by fluctuations in the motion of the valence electrons. Owing to very quick electron motion, the instantaneous dipole moment varies rapidly in time both as value and orientation. In the case of high symmetrical molecules, the average of the instantaneous dipole moment is null; this kind of molecules are considered neutral and nonpolar from the electric point of view.

In condensed media, the electric field created by the instantaneous electric dipole moment of a molecule acts on the neighboring molecules, modifying their dipole moments by induction, polarization, orientation, and so on.

Between nonpolar molecules act dispersion forces whose energies depend both on the molecular ionization potential and polarizability. The dispersion forces arising in a nonpolar molecule due to surrounding temporary dipole moments are the weakest attractive intermolecular forces. The weak interaction between the instantaneous dipole and induced dipole influences the potential energy of the dispersion forces affected by the solute polarizability and also by the molecular mass and shape. The dispersion forces become stronger when the polarizability increases. They also increase with the molecular mass and in the case of the elongated molecules.

All instantaneous interactions influence the molecular energies both in ground and excited electronic states of the solute molecules. The difference between these energies determines the spectral shift of the electronic absorption band related to its position in the solute gaseous phase. This difference depends on the strength of the dispersion interactions of solute in the electronic states responsible for the absorption phenomena in nonpolar solutions.

1.2 THEORETICAL BACKGROUND

The existence of the dispersive forces between the nonpolar molecules cannot be explained by classical theories. The first scientist who explained the dispersive interactions in a satisfactory theory was London (Murgulescu and Sahini 1978, Hodges and Stone 2000) in 1933. He has shown that, when two nonpolar molecules u and v are at a distance r_{uv} small enough that the long-range forces do not tend to zero, their instantaneous dipole moments μ_u and μ_v interact by attractive forces. The electric field around them modifies the temporary distribution of the valence electronic cloud in each molecule, by inductive actions. The new induced dipole moments have opposite senses and determine an intensification of the attractive forces between the molecules u and v.

The intensity of the dispersive forces varies in function on both intermolecular distance and polarizability of the spectrally active (solute) molecules in electronic states participating to the absorption phenomenon (Reichardt 2003, Parasegian 2006).

In London theory, the dispersive interaction energy w_{uv} between two molecules u and v is expressed by

$$w_{uv} = -\frac{3}{2}\frac{h\nu_{0u}h\nu_{0v}}{h\nu_{0u}+h\nu_{0v}}\frac{\alpha(u)\alpha(v)}{r_{uv}^6} \tag{1.1}$$

The parameters from relation (1.1) have the following significance:

w_{uv} is the energy due to dispersive interactions in the isolated molecular pair $u - v$;

$h\nu_{0u} = I(u)$ and $h\nu_{0v} = I(v)$ are ionization potentials of the two interacting molecules determined by the highest frequencies ν_{0u} and ν_{0v} from the molecular electronic absorption spectra;

$\alpha(u)$ and $\alpha(v)$ are the molecular polarizabilities of the molecules u and v, respectively;

r_{uv} is the intermolecular distance.

Relation (1.1) expresses the dispersion energy for one isolated molecular pair, corresponding to the gaseous phase.

In order to obtain the total energy due to dispersive interactions in the condensed phase, the dispersion energy is considered to be additive; and the solution is composed from all molecular pairs possible to be made between one solute and the solvent molecules; then, the energetic contribution of all pairs is finally summarized (Bakhshiev 1972).

In the absorption process, two electronic states are implied. They are characterized by different values of the molecular ionization potential and polarizability. Even the solvent keeps its electronic state (the solvent does not absorb in the spectral range in which the electronic absorption band of the solute is recorded), the dispersive energies in the electronic states participating to absorption phenomenon are different; their difference determines the spectral shift of the electronic absorption band recorded in solution, compared to the solute gaseous phase.

The energy of dispersive forces in a given nonpolar solvent can be estimated from the spectral shifts of the electronic absorption band relative to its position in the gaseous phase (Gerschel 1995, Lide 2008).

The position of the electronic absorption band in wavelength (wavenumber or frequency) scale is determined by the difference of the two solvation energies corresponding to the electronic states participating in the absorption process.

If the absorption takes place between the ground (g) and excited (e) electronic states (Figure 1.1), one can write

$$hc\bar{v}_l = hc\bar{v}_0 + (\Delta W_e - \Delta W_g) \tag{1.2}$$

The notations in relation (1.2) correspond to Figure 1.1 $\bar{v}$ is the wavenumber in the maximum of the electronic absorption band in solution ($\bar{v}_l$) and in gaseous phase ($\bar{v}_0$).

The dispersive forces are of attractive type. Consequently, the solvation energy due to this kind of interactions is a negative quantity which causes the solute stabilization.

The difference $\Delta W_e - \Delta W_g$ between the solvation energies in the two states participating in the absorption process measures the spectral shift of the electronic absorption band related to its position in the solute gaseous phase (Figure 1.1).

The studied solutions consist of very small amounts of solute (10^{-3}–10^{-5} mol/L) in a given solvent. In this condition, the interactions solute–solute can be neglected due to high distances between the spectrally active molecules. Only the dispersive interactions between one nonpolar solute molecule and the surrounding nonpolar solvent molecules contribute to the solvation energies.

The binary solution can be considered as being a homogeneous, continuous, and polarizable medium (McRae 1956, Bakhshiev 1972) with refractive index n, and one

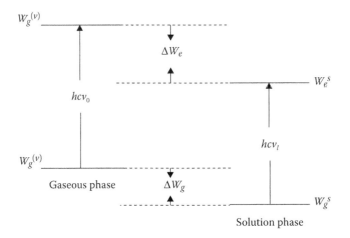

FIGURE 1.1   Energy levels of the spectrally active molecule when passing from gaseous phase to binary solution; W_e and W_g—electronic energies in excited and ground states, when the solute is in gaseous (v) or in solution (s); ΔW_e and ΔW_g—solvation energies of the solute in excited and ground states, respectively.

obtains the total solvation energy caused by dispersion interactions, by adding the energies corresponding to all molecular pairs solute–solvent, as it follows:

$$\Delta W = -\frac{3}{2}\frac{I(u)I(v)}{I(u)+I(v)}\frac{\alpha(u)}{r^3(u)}\cdot\frac{n^2-1}{n^2+2} \tag{1.3}$$

In relation (1.3), the parameters corresponding to the solute molecule are indexed by (u) and those to the solvent molecules by (v). The dispersion function (Murgulescu and Sahini 1978, Reichardt 2003) of the solvent is given by

$$\frac{\alpha(v)}{r(v)^3}=\frac{n^2-1}{n^2+2}=f(n)$$

From Equations 1.2 and 1.3, it results in

$$hc\bar{\nu}_l = hc\bar{\nu}_0 + \frac{3}{2}\frac{I_g(v)}{r^3(u)}\left[\frac{I_g(u)\alpha_g(u)}{I_g(v)+I_g(u)}-\frac{(I_g(u)-hc\bar{\nu}_l)\alpha_e(u)}{I_g(v)+I_g(u)-hc\bar{\nu}_l}\right]f(n) \tag{1.4}$$

The relation between the ionization potentials in ground and excited states was used (Eliashevici 1966):

$$I_e(u)=I_g(u)-hc\bar{\nu}_l \tag{1.5}$$

From Equation 1.5, it results in a linear dependence of type (1.6) between the wavenumber in the maximum of the absorption band recorded in binary solutions and the solvent dispersion function $f(n)$.

$$\bar{v}(\text{cm}^{-1}) = \bar{v}_0(\text{cm}^{-1}) + mf(n) \tag{1.6}$$

The slope of this line is given by

$$m = \frac{3}{2} \frac{I_g(v)}{r^3(u)} \left[\frac{I_g(u)\alpha_g(u)}{I_g(v) + I_g(u)} - \frac{(I_g(u) - hc\bar{v}_l)\alpha_e(u)}{I_g(v) + I_g(u) - hc\bar{v}_l} \right] \tag{1.7}$$

The mathematical expression of the slope m contains the solute-excited state polarizability $\alpha_e(u)$, as an unknown parameter. A series of molecular parameters from Equation 1.8 can be computed by using quantum chemical programs, or can be experimentally determined. One obtains

$$\alpha_e(u) = \frac{I_g(v) + I_g(u) - hc\bar{v}_l}{I_g(u) - hc\bar{v}_l} \frac{3}{2} \frac{I_g(v)}{r^3(u)} \left[\frac{I_g(u)\alpha_g(u)}{I_g(v) + I_g(u)} - \frac{2mr^3(u)}{3I_g(v)} \right] \tag{1.8}$$

Relation (1.8) shows that the excited state polarizability of the spectrally active molecule can be estimated when the ionization potentials, the molecular radius, and the ground state polarizability are computed and the wavenumber in the maximum of the absorption band is experimentally determined in nonpolar solutions. The ionization potentials of solute (u) and solvent (v) are usually known or can be estimated by using quantum mechanical programs; the ground state polarizability can also be computed; the wavenumber in the maximum of the electronic absorption band is experimentally measured (see Tables 1.1 and 1.2) and the slope m of the line in plots $\bar{v}_l(\text{cm}^{-1})$ versus $f(n)$ (see Figure 1.2) can be graphically or statistically estimated (Dorohoi et al. 2009).

TABLE 1.1
Wavenumber $\tilde{v}$ (cm^{-1}) in the Maximum of Electronic Absorption Band of Benzene (Be), Naphthalene (Naph), Anthracene (An), and Tetracene (Tetr) in Nonpolar Solvents Characterized by Refractive Index n and Ionization Potential $I_g(v)$

Solvent	n	$I_g(v)$ (eV)	$\bar{v}_l$ (cm^{-1}) Be	Naph	An	Tetr
Pentane	1.358	10.53	39,280	32,180	26,720	21,382
Hexane	1.375	10.13	39,280	32,174	26,680	21,336
Heptane	1.387	10.10	39,260	32,169	26,638	21,331
n-Octane	1.398	9.82	39,270	32,165	26,643	21,310
Cyclopentane	1.407	10.56	39,260	32,165	26,630	21,272
Decane	1.411	10.20	39,265	32,160	26,625	21,264
Dodecane	1.421	9.60	39,260	32,160	26,588	21,220
Cyclohexane	1.427	9.80	39,255	32,154	26,596	21,228

TABLE 1.2
Wavenumber $\bar{\nu}_l$ (cm^{-1}) in the Maximum of the Electronic Absorption Band of Phenanthrene (Phena), Chlorobenzene (ClBe), Nitrobenzene (NiBe), 9,10-Dinitroanthracene (9,10-DNiAn), and 1,6-Diphenyl-1,3,5-Hexatriene (DPH) in Nonpolar Solvents

	$\bar{\nu}_l$ (cm^{-1})				DPH		
Solvent	Phena	ClBe	NiBe	9,10-DNiAn	Band 1	Band 2	Band 3
Pentane	34,525	37,790	39,790	25,450	27,008	28,502	29,946
Hexane	34,415	37,780	39,778	25,400	27,064	28,490	29,940
Heptane	24,280	37,770	39,717	25,380	26,954	28,409	29,851
n-Octane	24,225	37,765	39,521	25,300	26,826	28,280	29,790
Cyclopentane	34,180	37,750	39,543	25,325	26,780	28,235	29,650
Decane	34,110	37,740	39,486	25,300	26,770	28,210	29,632
Dodecane	34,050	37,700	39,572	25,295	26,710	28,164	29,610
Cyclohexane	34,052	37,705	39,499	25,270	26,810	28,249	29,762

The solute polarization by the solvent permanent dipole moments (Baur and Nicol 1966) influences the spectral shift relative to the gaseous phase of the solute, especially in solutions containing nonpolar but polarizable solute and a few polar, or polar and nonviscous solvent. So, one term due to the statistical fluctuations of the medium depending on temperature and the solution electric permittivity (Baur and Nicol 1966, Dimitriu 2009, Gheorghies et al. 2010) must be added to the term describing the effect of dispersion forces on the spectral shifts, as follows:

$$ hc\bar{\nu}_l = hc\bar{\nu}_0 + mf(n) + C\frac{(2\varepsilon+1)(\varepsilon-1)}{\varepsilon} \tag{1.9} $$

with

$$ C = -\frac{12kT}{R^3}\ln^2\left(\frac{R}{a}\right)(\alpha_e(u)-\alpha_g(u)) \tag{1.10} $$

As it was established in literature (Baur and Nicol 1966), the constant C depends on the solute polarizability in the electronic states responsible for the electronic transition, on temperature T, on the solvent shell radius R, and also on the molecular radius r. The value of the solvent shell radius R must be sufficiently large so that the dielectric properties within the cavity should be approximately equal to those of the bulk solution. Baur and Nicol consider R being of the same order as the mean distance between the solute molecules in solution, which usually is in the range of 10–20 Å. The value of the coefficient C results from the statistical analysis of the spectral data.

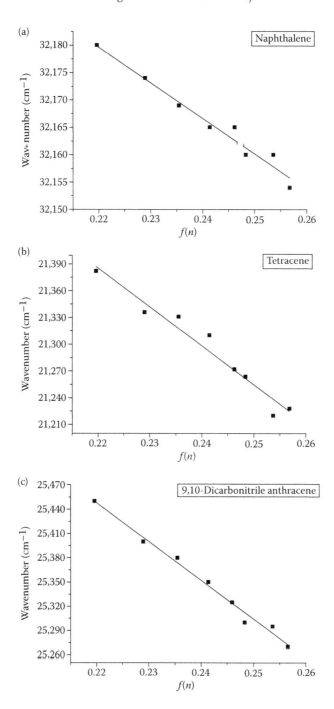

FIGURE 1.2 Wavenumber in the maximum of the electronic absorption band versus dispersive function $f(n)$ of solvent, for naphthalene (a), tetracene (b) and 9,10-dicarbonitrile anthracene (c), respectively.

The theory developed by T. Abe (Abe 1965) permits to estimate the excited states dipole moment and polarizability of a spectrally active molecule from the solvato-chromic study. The final formula obtained by T. Abe is of the type

$$\mu_e^2(u) - \mu_g^2(u) + \alpha_e(u)a - b \qquad (1.11)$$

In Equation 1.11, a and b are two parameters depending on the solvent and solute characteristics as it follows:

$$a = \frac{\frac{3}{2}\left\{(n^2-1)I_g\dfrac{I_g(u)-hc\nu_l}{I_g+I_g(u)-hc\nu_l} + 2kT\dfrac{(\varepsilon-n^2)(2\varepsilon+n^2)}{\varepsilon(n^2+2)}\right\}}{2\dfrac{(\varepsilon-n^2)(2\varepsilon+n^2)}{\varepsilon(n^2+2)}+n^2-1} \qquad (1.12)$$

$$b = \frac{-\dfrac{\nu_l-\nu_\nu}{C}+\dfrac{9}{8\pi N_A}\dfrac{M}{\rho}\dfrac{n^2-1}{n^2+2}I_g\alpha_g(u)\dfrac{I_g(u)}{I_g+I_g(u)}+\dfrac{9kT}{4\pi N_A}\dfrac{M}{\rho}\dfrac{(\varepsilon_V-n^2)(2\varepsilon+n^2)}{\varepsilon(n^2+2)^2}\alpha_g(u)}{\dfrac{3}{2\pi N_A}\cdot\dfrac{M}{\rho}\cdot\dfrac{(\varepsilon-n^2)(2\varepsilon+n^2)}{\varepsilon(n^2+2)^2}+\dfrac{3}{4\pi N_A}\cdot\dfrac{M}{\rho}\cdot\dfrac{n^2-1}{n^2+2}}$$

$$(1.13)$$

Relations (1.11) through (1.13) permit to establish the applicability of T. Abe model to the studied spectrally active molecules and also to estimate in the limits of this cell model, the values of the excited state dipole moment and polarizability, if the dipole moment of the studied molecule is computed by quantum chemical proce-dures (Abe 1966, Abe et al. 1966).

Let us analyze some results obtained in literature, referring to the spectrally active nonpolar molecules or to some molecules with relatively small dipole moments solved in small amounts in homogeneous liquids.

1.3   RESULTS AND DISCUSSIONS

In nonpolar solutions (made in nonpolar solvents with nonpolar spectrally active molecules in very low concentrations), the spectral shifts of the electronic absorp-tion bands relative to the gaseous phase are prevalently caused by dispersion forces (Dorohoi and Dimitriu 2001, 2007) and are generally described by a linear depen-dence between the wavenumber in the band maximum and the dispersion function with slope from relation (1.7).

In order to sustain this affirmation, the electronic absorption spectra of some ben-zene derivatives were recorded on Specord UV-VIS spectrophotometer Carl Zeiss Jena with data acquisition system in a series of nonpolar solvents (see Tables 1.1 and 1.2). The solvents, achieved from Merck Company, were spectrally grade and used without purification. The measurements were five times repeated and the average value of the wavenumber in the maximum in the electronic absorption band was listed in Tables 1.1 and 1.2. The obtained values are in a good accordance with those reported previously (Nicol et al. 1968, Macovei 1975, 1976a,b, Pop et al. 1978, Strat and Strat 2000).

Let us analyze some results obtained for a series of benzene derivatives, nonpolar but polarizable solutes: benzene, naphthalene, anthracene, tetracene, and phenanthrene (see Tables 1.1 and 1.2).

A quantum mechanical analysis of the studied benzene derivatives has been made using Spartan'14 programs. The results are listed in Table 1.3 (Dimitriu et al. 2008a,b).

Considering the previous results reported for L_{1a} and L_{1b} electronic absorption bands of the substances from this series (Macovei 1975), and the results from Table 1.3, one can say that the ground state polarizability of the benzene derivatives increases with the number of the rings attached to the benzene cycle, excepting phenanthrene. An increase in the ground state polarizability can also be reported when some substituents are added to the benzene cycle.

The wavenumbers in the maximum of the electronic absorption bands were correlated with the solvent dispersion $f(n)$ function (see Figure 1.2). The correlation characteristics, obtained by using Origin 8.0, are listed in Table 1.4. The slope of the line $v_l = v_0 + mf(n)$ is negative in Table 1.4, for all studied compounds, showing in accordance with relation (1.7), the increase in the molecular polarizability by excitation, in the absorption process.

The dependence expressed in Equation 1.6 is plotted in Figure 1.2 for some molecules of Tables 1.1 and 1.2.

As it results from Equation 1.7, the slope of the lines $\bar{v}_l(cm^{-1})$ versus $f(n)$ directly depends on the excited state polarizability of the spectrally active molecule. In the case of nonpolar and nonpolarizable benzene derivatives, the slope m of line (1.7) increases with increasing the benzene ring number of the solute molecule.

TABLE 1.3
Results of Quantum Chemical Analysis Made with Spartan'14

Solute Molecule	$\mu_g(u)$ (D)	$I_g(u)$ (eV)	$\alpha_g(u)$ (Å³)	Volume (Å³)	Surface Area (Å²)
Be	0	9.75	47.03	99.15	115.07
Naph	0	8.84	51.58	150.35	161.90
An	0	8.25	56.01	201.57	208.68
Tetra	0	7.87	60.34	252.78	255.44
Phena	0	8.74	55.74	200.92	205.81
9,10-DiNiAn	0	8.87	50.11	238.45	241.39
ClBe	0	9.38	48.26	112.32	121.40
NiBe	0	10.62	48.9	121.4	143.54
DPH[a]	0	8.13	30.53	820.18	516.68

[a] Computed with HyperChem.

TABLE 1.4
Results of Statistical Analysis of the Spectral Data ($v_I = v_0 + mf(n)$)
and the Excited State Polarizability α_e

Solute Molecule	Slope m (cm^{-1})	Cut at Origin v_0 (cm^{-1})	α_e (Å^3)	$\Delta\alpha$ (Å^3)
Be	-642 ± 73	$39{,}422 \pm 18$	70.80	23.78
Naph	-647 ± 51	$32{,}322 \pm 12$	83.27	31.69
An	-3402 ± 210	$27{,}463 \pm 51$	78.33	22.32
Tetra	-4386 ± 346	$22{,}351 \pm 84$	79.90	19.59
Phena	$-13{,}392 \pm 770$	$37{,}460 \pm 186$	97.62	36.52
9,10-DiNiAn	-4772 ± 225	$26{,}498 \pm 54$	78.89	19.78
ClBe	-2238 ± 271	$38{,}292 \pm 66$	73.70	25.44
NiBe	-8417 ± 469	$41{,}644 \pm 114$	71.90	23.00
DPH-Band 1	-8636 ± 567	$28{,}907 \pm 142$	65.96	35.43
DPH-Band 2	-8286 ± 495	$30{,}125 \pm 125$	66.67	36.14
DPH-Band 3	-7417 ± 689	$31{,}465 \pm 173$	66.04	35.51

The excited state polarizability of the benzene derivatives increases with the number of the bonded benzene rings. Consequently, the increase in polarizability in the excited state can be correlated with the number of the conjugated π bonds.

Some physical and chemical characteristics obtained in molecular modeling by HyperChem 8.06 for 1,6-diphenyl-1,3,5-hexatriene (DPH), a nonpolar and polarizable molecule due to its high number of π-conjugated chemical bonds are listed in Table 1.3. A linear dependence of the wavenumber in the maximum of the vibronic bands of DPH in function of the dispersion function was reported (Hurjui et al. 2013) for 34 solvents. The data reported here refer only to the nonpolar solvents from Tables 1.1 and 1.2.

DPH molecule is a nonpolar, but polarizable one and the dispersion forces in its solutions are prevalent in nonpolar solvents (see Tables 1.1 and 1.2), as it was established (Hurjui et al. 2013). For the three vibronic bands of DPH in nonpolar solvents from Tables 1.1 and 1.2, the excited state polarizability was estimated as being near 66 Å^3 (see Table 1.4).

Some polar solutes, such as ClBe and NiBe, also show a linear dependence of the wavenumber in the maximum of the electronic absorption band and the dispersion function. Some other studies regarding the solvent influence on the electronic absorption band of these compounds emphasized the linear dependence between the wavenumber in the maximum of the electronic absorption band and the solvent dispersion function (Macovei 1976a,b).

The results given in Tables 1.1 through 1.3 show that the polar molecules in the nonpolar and nonpolarizable solvents are very sensitive to the dispersion forces, especially nitrobenzene, for which the studied band is one with charge transfer from the benzene cycle toward the nitro group (Tanakaa et al. 2003). The data referring to quantum mechanical calculation and to statistical analysis of the spectral data from Tables 1.1 and 1.2 are given in Tables 1.3 and 1.4.

A linear dependence between the $\bar{v}_l$ (cm^{-1}) and $f(n)$, with statistic parameters listed in Table 1.4, was also obtained for 9,10-di-carbonitrile anthracene. Its physical–chemical parameters computed by Spartan'14 are listed in Table 1.3.

The ionization potentials of solute (u) and solvent (v) are usually known or can be estimated by using quantum mechanical programs; the ground state polarizability can also be computed; the wavenumber in the maximum of the electronic absorption band is experimentally measured (see Tables 1.1 and 1.2) and the slope m of line in the plots $\bar{v}_l$ (cm^{-1}) versus $f(n)$ can be graphically or statistically estimated on the basis of the experimental data.

The excited state polarizability of solute, $\alpha_e(u)$, estimated in each solvent using relation (1.8) shows its small dependence on the solvent nature.

The inconvenience of this method is the fact that some quantum chemical computations are made on the isolated molecules and the programs from estimating molecular parameters in the electronic ground state are developed for isolated molecules (in the gaseous phase). Up to now, the quantum mechanical calculations for solutions were realized for a few numbers of molecules in a small number of solvents. The information referring to the excited state polarizability or dipole moments will enrich the database and can contribute to develop new quantum mechanical programs for the molecular-excited states.

In solutions achieved in a great variety of solvents, containing nonpolar solute molecules or solute molecules with small dipole moments, the dependencies between the wavenumbers in the maximum of the absorption bands and the macroscopic parameters (n and ε) of the solvents show the contribution of other universal interactions to the spectral shifts. Baur and Nicol (Baur and Nicol 1966, Nicol et al. 1968, Gheorghies et al. 2010) underlined the importance of the thermal fluctuations in polar solutions and obtained a term which can describe this kind of interactions.

By using the theory developed by Baur and Nicol, the excited state polarizability of some amino-nitro-benzo derivatives from the solvent shifts recorded in electronic absorption spectra was estimated (Dorohoi et al. 2009). The results are listed in Tables 1.5 and 1.6.

TABLE 1.5
Spectral Data and Solvent Parameters of the Studied Nitro-Anilines

Solvent	N	$I_g(v)$ (eV)	$\bar{v}_l$ (cm^{-1}) 2-Ni A	4A-3NiP	3A-4NiP
Benzene	1.539	9.24	25,440	23,000	26,390
o-Xylene	1.537	8.56	25,570	23,040	26,350
Diethyl ether	1.352	9.53	25,250	22,680	26,180
Chloroform	1.360	11.42	25,320	23,040	26,110
Dichloromethane	1.430	11.35	25,120	22,990	26,040
1,2-Dichloroethane	1.454	11.06	25,190	22,990	26,110
Ethanol	1.369	10.70	24,690	21,930	25,250
Methanol	1.335	10.84	24,750	22,030	25,510

TABLE 1.6

Electro-Optical Parameters of the Studied Molecules Computed by HyperChem (First Three Columns) and Results of the Spectral Data Analysis of the Excited State Polarizability

Molecule	μ_g (D)	I_g (eV)	$10^{24} \cdot \alpha_g$ (cm³)	$\bar{\nu}_o \pm \Delta\bar{\nu}_o$ (cm⁻¹)	$C \pm \Delta C$ (cm⁻¹)	$10^{24} \cdot \alpha_e$ (cm³)
2-NiA	4.8	9.23	13.6	$22{,}322 \pm 46$	-14.2 ± 1.1	18.8
4A-3NiP	3.8	8.82	14.3	$22{,}555 \pm 64$	-13.7 ± 1.3	19.3
3A-4NiP	4.3	9.49	14.3	$26{,}293 \pm 57$	-18.3 ± 1.4	20.9

Note: The excited state polarizability of the studied aniline derivatives is higher than their ground state polarizability.

In the case of 2NiA (2-nitro-aniline), 4A3NiP (4-amino-3-nitro-phenol), and 3A4NiP (3-amino-4-nitro-phenol), the values estimated by using the thermal fluctuations in polar solutions (Dimitriu 2009) for the excited state polarizability of the studied compounds increased from 2NiA to 3A4NiP (see Table 1.6). Some spectral data were repeated and appropriate values as in Dimitriu (2009) were obtained with spectrally grade solvents. The obtained results are concordant with literature data (Moran and Kelleya 2001, Dorohoi et al. 2009).

Relation (1.9) is accomplished for $m = 0$ for the nitro-aniline derivatives proving the influence of the statistical fluctuations of the medium depending on temperature and the solution electric permittivity (Nicol et al. 1968, Dimitriu 2009). The data resulting from statistical analysis applied to a great number of solvents (Dimitriu 2009) are listed in Table 1.6 in which the first three columns contain the values of dipole moment, ionization potential, and ground state polarizability of the studied molecules in their ground electric state, computed by HyperChem.

Takehiro Abe theory developed for a simple liquid in which the solute substance is solved in very small quantities permits to compute, based on the solvatochromic study, the excites state values both of the dipole moment and polarizability.

Let us apply the model proposed by T. Abe for estimating the excited state polarizability for the vibronic bands of anthracene. The obtained results are shown in Table 1.7 (wavenumbers in cm⁻¹) and in Table 1.8 (excited state polarizability and dipole moment of anthracene computed for the three vibronic bands). The model allows estimating the difference of the square dipole moments in the electronic states participating in the absorption process (Dorohoi and Dimitriu 2001, Dimitriu et al. 2007). Figure 1.3 shows the linear fit of the dependence between the parameters a and b in Abe's model for the three vibronic bands of anthracene.

The excited state polarizability of anthracene is higher than its ground state polarizability $\alpha_g(u) = 56$ Å³ and decreases with the vibration number. By the photon absorption, the dipole moment of anthracene increases in all vibronic states responsible for the visible vibronic band appearance (Dorohoi and Dimitriu 2001).

The T. Abe model was also applied to the vibronic components of some cyclo-adducts of benzo-[f]-quinoline, with the structure from Figure 1.4 for which the

TABLE 1.7

Wavenumbers $\bar{\nu}$ (cm^{-1}) in the Maxima of the Vibronic Bands of Anthracene

Solvent	Band 1	Band 2	Band 3
Cyclohexane	26,500	27,940	29,400
Carbon tetrachloride	26,335	27,740	29,160
Mesitylene	26,375	27,810	29,230
Toluene	26,375	27,680	29,425
o-Xylene	26,355	27,810	29,215
m-Xylene	26,375	27,820	29,200
p-Xylene	26,315	27,745	29,170
Benzene	26,290	27,700	29,170
Methyl acetate	26,540	27,980	29,400
Ethanol	26,540	27,980	29,400
Acetone	26,480	27,920	29,340
Ethyl acetate	26,500	27,935	29,360
Pyridine	26,230	27,670	29,080
Propionic acid	26,585	28,000	29,425
Chlorobenzene	26,350	27,660	29,130
2-Butanone	26,500	27,915	29,340
Anisole	26,290	27,730	29,145

TABLE 1.8

Excited State Polarizability, Difference of the Squares of Dipole Moments in Ground and Excited States, Correlation Coefficient, R, Ground State $\mu_g(u)$, and Excited State $\mu_e(u)$ Dipole Moments for the Vibronic Bands of Anthracene

Band	$\alpha_g(u)$ (10^{-24} cm^3)	$\mu_e^2(u) - \mu_g^2(u)$ (D^2)	R	$\mu_g(u)$ (D)	$\mu_e(u)$ (D)
1	59.09	0.52	0.997	0	0.72
2	58.88	0.77	0.998	0	0.88
3	55.18	1.10	0.998	0	1.05

wavenumbers in the maxima are listed in Tables 1.9 through 1.11. These compounds were previously studied from the solvatochromic point of view and the contribution of the dispersion forces to the spectral shifts of the $\pi \rightarrow \pi^*$ transitions was established on the basis of dependence (1.6) (Dorohoi et al. 1980, Dorohoi and Iancu 1981).

The applicability of T. Abe model to the vibronic bands of cycloadducts C1–C3 results from the fits of parameters b and a (Figure 1.5).

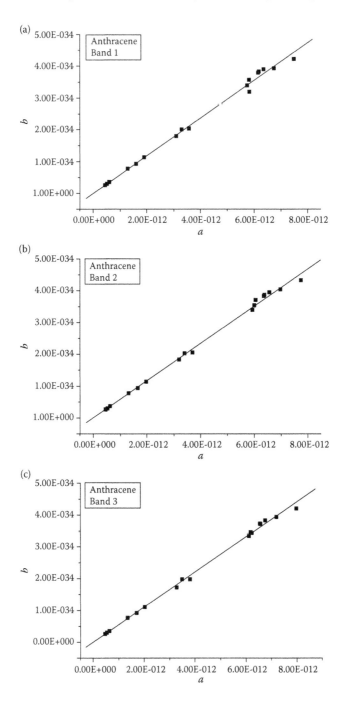

FIGURE 1.3   Linear fit (according to Equation 1.11) of the experimental data corresponding to the three spectral bands of anthracene.

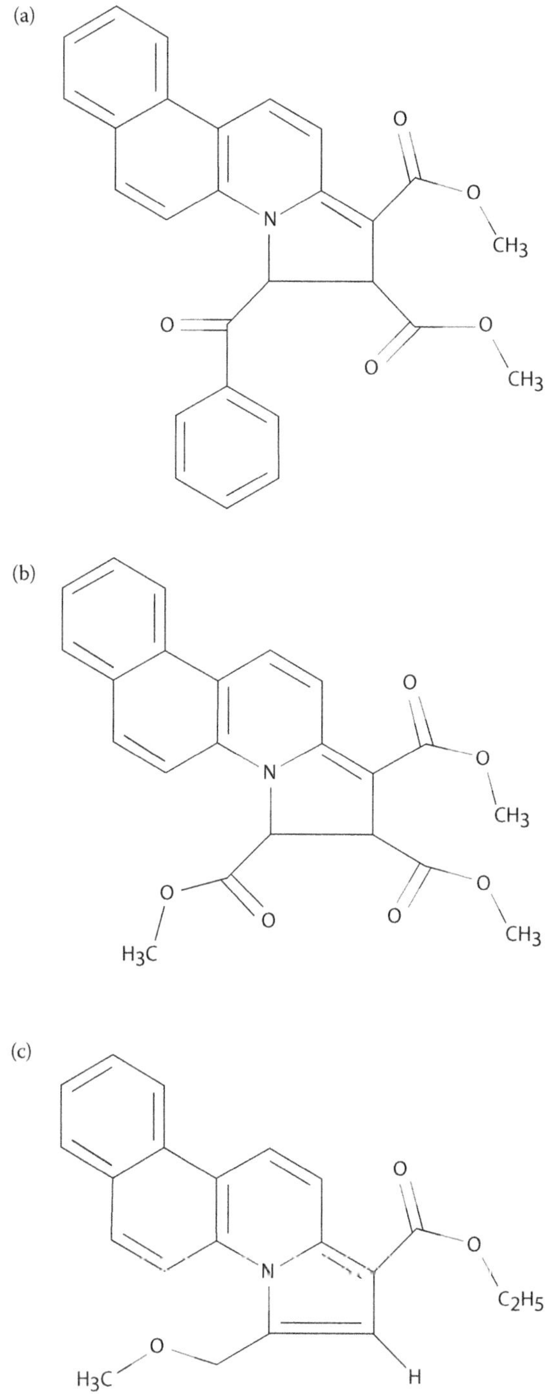

FIGURE 1.4 Structural features of the studied cycloadducts.

TABLE 1.9
Wavenumbers, $\bar{v}\,(cm^{-1})$, of the Vibronic Bands of C1

Solvent	Band 1	Band 2	Band 3	Band 4	Band 5	Band 6
Dichloroethane	18,960	20,160	21,440	22,800	27,280	28,240
Chloroform	19,000	20,400	21,640	22,780	27,520	28,600
Toluene	18,850	19,920	21,240	22,580	26,920	28,040
o-Xylene	18,500	19,960	21,360	22,900	26,750	28,100
Phenyl chloride	18,750	20,080	21,420	22,640	27,120	28,200
Benzene	18,660	20,080	21,320	22,920	27,040	28,100
Anisole	18,620	20,040	21,400	22,600	27,180	28,300

TABLE 1.10
Wavenumbers, $\bar{v}\,(cm^{-1})$, of the Vibronic Bands of C2

Solvent	Band 1	Band 2	Band 3	Band 4	Band 5	Band 6
Dichloroethane	19,240	20,420	21,700	22,940	27,200	28,200
Carbon tetrachloride	18,920	20,240	21,540	22,820	22,660	27,880
Toluene	18,900	20,320	21,580	22,900	26,940	28,140
p-Xylene	18,920	20,480	21,620	23,000	26,900	28,100
o-Xylene	19,000	20,300	21,620	22,880	–	–
Phenyl chloride	19,000	20,280	21,580	22,740	26,940	28,000
Benzene	19,000	20,340	21,640	22,900	26,900	28,080
Anisole	19,000	20,300	21,580	23,160	27,040	28,140

TABLE 1.11
Wavenumbers, $\bar{v}\,(cm^{-1})$, of the Vibronic Bands of C3

Solvent	Band 1	Band 2	Band 3	Band 4	Band 5
Dichloroethane	25,540	26,860	28,160	–	32,320
Toluene	25,440	26,740	28,020	29,480	32,120
o-Xylene	25,420	26,710	28,120	29,600	–
Phenyl chloride	25,360	26,670	28,080	29,500	32,060
Carbon tetrachloride	25,490	26,800	28,120	29,580	32,100
Ethyl acetate	25,700	27,020	28,280	29,680	32,480
Anisole	25,380	26,690	27,940	29,560	32,140

The values of the polarizability and of the dipole moment for various vibronic bands (Creanga et al. 2001) estimated by applying Abe model are listed in Tables 1.12 through 1.14.

The benzo-[f]-quinolinium cycloadducts C1–C3 were studied from the solvent influence on the electronic absorption bands point of view and the data were

TABLE 1.12

Excited State Polarizability and Dipole Moment, Estimated on the Basis of T. Abe Model for Each Vibronic Band of Cycloadduct C1

Band	$\alpha_e(u)$ (10^{-24} cm³)	$\mu_e^2(u) - \mu_g^2(u)$ (D²)	R	$\mu_g(u)$ (D)	$\mu_e(u)$ (D)
1	48.87	20.49	0.998	3.67	5.82
2	51.78	14.06	0.998	3.67	5.24
3	58.36	12.33	0.998	3.67	5.07
4	35.73	4.01	0.973	3.67	4.18
5	134.11	1.37	0.992	4.67	3.89
6	143.88	−7.75	0.995	3.67	2.38

TABLE 1.13

Excited State Polarizability and Dipole Moment, Corresponding to Each Vibronic Band of Cycloadduct C2, Estimated on the Basis of T. Abe Model

Band	$\alpha_e(u)$ (10^{-24} cm³)	$\mu_e^2(u) - \mu_g^2(u)$ (D²)	R	$\mu_g(u)$ (D)	$\mu_e(u)$ (D)
1	43.45	17.18	0.998	4.10	5.83
2	44.34	8.17	0.998	4.10	4.99
3	44.01	4.00	0.998	4.10	4.56
4	49.67	3.38	0.997	4.10	4.49
5	122.76	3.60	0.989	4.10	4.51
6	97.30	0.51	0.995	4.10	4.16

TABLE 1.14

Excited State Polarizability and Dipole Moment Corresponding to Each Vibronic Band of Cycloadduct C3, Estimated on the Basis of T. Abe Model

Band	$\alpha_e(u)$ (10^{-24} cm³)	$\mu_e^2(u) - \mu_g^2(u)$ (D²)	R	$\mu_g(u)$ (D)	$\mu_e(u)$ (D)
1	148.22	4.58	0.999	2.34	3.17
2	150.85	5.02	0.999	2.34	3.24
3	141.64	2.87	0.998	2.34	2.88
4	144.13	3.42	0.999	2.34	2.98
5	150.74	19.42	0.999	2.34	4.98

interpreted by using the existent theories about the dispersion forces in liquid nonpolar solutions (Dorohoi et al. 1980, Dorohoi and Iancu 1981). The wavenumbers in the maxima of the electronic absorption bands of the studied cycloadducts due to $\pi \rightarrow \pi^*$ transitions satisfy the following equations:

For C1:

$$\overline{v}\,(\text{cm}^{-1}) = 30{,}197 - 7299\,f(n) \quad \text{(band 4)}; \quad \alpha_e(u) = 112.24\text{Å}^3$$
$$\overline{v}\,(\text{cm}^{-1}) = 22{,}814 - 10{,}193\,f(n) \quad \text{(band 5)}; \quad \alpha_e(u) = 111.94\text{Å}^3 \tag{1.14}$$

For C2:

$$\overline{v}\,(\text{cm}^{-1}) = 29{,}240 - 3928\,f(n) \quad \text{(band 4)}; \quad \alpha_e(u) = 101.53\text{Å}^3$$
$$\overline{v}\,(\text{cm}^{-1}) = 22{,}779 - 6437\,f(n) \quad \text{(band 5)}; \quad \alpha_e(u) = 104.92\text{Å}^3 \tag{1.15}$$

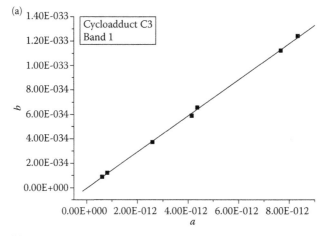

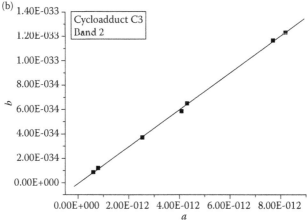

FIGURE 1.5   Linear fit of T. Abe parameters a and b for two vibronic bands of cycloadduct C3.

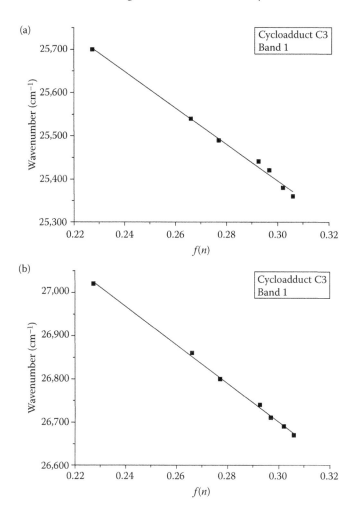

FIGURE 1.6   Linear fit of wavenumber in the maximum of visible absorption band and the dispersion function for two vibronic bands of cycloadduct C3.

For C3:

$$\bar{v}\,(\mathrm{cm}^{-1}) = 33{,}030 - 3020\,f(n) \quad (\text{band 1}); \quad \alpha_e(u) = 94.78\,\mathring{A}^3$$
$$\bar{v}\,(\mathrm{cm}^{-1}) = 30{,}520 - 3420\,f(n) \quad (\text{band 2}); \quad \alpha_e(u) = 96.99\,\mathring{A}^3$$
$$\bar{v}\,(\mathrm{cm}^{-1}) = 29{,}060 - 3550\,f(n) \quad (\text{band 3}); \quad \alpha_e(u) = 98.98\,\mathring{A}^3 \qquad (1.16)$$
$$\bar{v}\,(\mathrm{cm}^{-1}) = 28{,}040 - 4570\,f(n) \quad (\text{band 4}); \quad \alpha_e(u) = 101.93\,\mathring{A}^3$$
$$\bar{v}\,(\mathrm{cm}^{-1}) = 26{,}720 - 4600\,f(n) \quad (\text{band 5}); \quad \alpha_e(u) = 107.49\,\mathring{A}^3$$

In Figure 1.6, the above-mentioned dependences are illustrated for two bands of C3 cycloadduct.

By using the slopes of the lines described by the written above equations, the excited state polarizability corresponding to the studied cycloadducts in different vibronic states was estimated. Results are written for each vibronic band of cycloadducts in the last column of relations (1.14) through (1.16). They are of the same order of magnitude as those obtained based on T. Abe model.

The increase in the polarizability by excitation is higher for $\pi \to \pi^*$ transitions compared to $n \to \pi^*$ transitions. The values of the excited state polarizability are comparable for the three cycloadducts under the study. They are smaller than those obtained by using T. Abe model, probably due to the fact that in order to apply a formula of the dispersion functions, only to nonpolar solvents, the values of the slope m in relation (1.6) are affected by a small number of the nonpolar solvents used for determination.

1.4   CONCLUSIONS

The linear relation between the wavenumber $\bar{v}_l$ (cm^{-1}) in the maximum of the electronic absorption band of the studied compounds in nonpolar solvents and the dispersion function $f(n)$ shows that the dominant factor in nonpolar solutions is the dispersion interaction.

The red shifts due to dispersion interactions (such in the case of benzene derivatives, benzo[f]quinolinium cycloadducts, aniline derivatives, and so on) correspond to $\pi \to \pi^*$ transitions and indicate the increase in the molecular polarizability by the molecular excitation.

The increase in polarizability is higher for the bands situated near UV region.

The values of the excited state polarizability due to the dispersion forces are comparable when they are estimated by the three described methods.

REFERENCES

Abe, T. 1965. Theory of solvent effects on molecular electronic spectra: Frequency shifts. *Bull. Chem. Soc. Japan* 30(8):1314–1318.

Abe, T. 1966. The dipole moments and polarizabilities in the $n \to \pi^*$ excited state of acetone from spectral solvent shifts. *Bull. Chem. Soc. Japan* 39(5):936–939.

Abe, T., Amako, Y., Nishioka, T., and Azumi, H. 1966. The dipole moments and polarizabilities in the excited stat of naphthalene from spectral solvent shifts. *Bull. Chem. Soc. Japan* 39(4):845–846.

Bakhshiev, N.G. 1972. *Spectroscopia Mejmoleculiar'x Bzaimodeistvii*. Nauka, Leningrad.

Baur, M.E. and Nicol, M. 1966. Solvent stark effect and spectral shifts. *J. Chem. Phys.* 44(9):3337–3343.

Creanga, D.E., Dimitriu, D.G., and Dorohoi, D.O. 2001. Electric dipole moments of some benzo-[f]-quinolinium cycloadducts. *Stud. Univ. Babes Bolyai Phys.*, Special Issue PIM:337–342.

Dimitriu, M. 2009. *Metode spectrale si de modelare moleculara pentru estimarea unor parametri electro-optici ai moleculelor organice*. PIM, Iasi.

Dimitriu, M., Dimitriu, D.G., and Dorohoi, D.O. 2007. Polarizability and excited state dipole moments in the excited states of some cycloadducts. *Proceedings of the 6th International Conference of Balkan Physical Union*, Istanbul, Turkey, *AIP Proceedings* 899:245–246.

Dimitriu, M., Ivan, L.M., and Dorohoi, D.O. 2008a. Electro-optical parameters of some benzene derivatives obtained by molecular orbital calculations. *Rev. Chim. Bucur.* 59(2):216–219.

Dimitriu, M., Ivan, L.M., and Dorohoi, D.O. 2008b. Electro-optical parameters of some chlorobenzene derivatives obtained by molecular orbital calculations. *Rom. J. Phys.* 53(1–2):75–80.

Dorohoi, D.O., Airinei, A., and Dimitriu, M. 2009. Intermolecular interactions in solutions of some amino-nitro-benzene derivatives, studied by spectral means. *Spectrochim. Acta A: Mol. Biomol. Spectrosc.* 73:257–262.

Dorohoi, D.O. and Dimitriu, D.G. 2001. Microscopic parameters in the excited state of some anthracene derivatives. *Stud. Univ. Babes Bolyai Phys.*, Special Issue PIM:332–336.

Dorohoi, D.O. and Dimitriu, M. 2007. Intermolecular interactions in non-polar liquids evidenced by spectral means. *Rev. Chim. Bucur.* 58(11):1060–1063.

Dorohoi, D.O., Guyre Rotariuc, M., and Iancu, D. 1980. Intermolecular interactions in some cycloadducts solutions I. *Analele Stiint. Univ. Alexandru Ioan Cuza Sectia Ib Fiz.* XXVI:71–76.

Dorohoi, D.O. and Iancu, D. 1981. Intermolecular interactions in some cycloadducts solutions II. *Analele Stiint. Univ. Alexandru Ioan Cuza Sectia Ib Fiz.* XXVII:49–54.

Eliashevici, M.A. 1966. *Atomic and Molecular Spectroscopy (in Romanian)*. Editura Academiei RSR, Bucuresti.

Gerschel, A. 1995. *Liaisons Intermoléculaire*. Inter science Editions et CNRS Editions, Paris.

Gheorghies, L.V., Dimitriu, M., Filip, E., and Dorohoi, D.O. 2010. Universal interactions in binary solutions studied by spectral means. *Rom. J. Phys.* 55(1–2):103–109.

Hodges, M.P. and Stone, A.J. 2000. A new representation of the dispersion interactions. *Mol. Phys.* 98:275.

Hurjui, I., Ivan, L.M., and Dorohoi, D.O. 2013. Solvent influence on the electronic absorption spectra (EAS) of 1,6-diphenyl-1,3,5-hexatriene (DPH). *Spectrochim. Acta A: Mol. Biomol. Spectrosc.* 102:219–225.

Lide, D.R. (Ed.). 2008–2009. *CRC Handbook of Chemistry and Physics*, 89th edition. CRC Press, Taylor & Francis Group, Boca Raton.

Macovei, V. 1975. Spectral solvent effects in aromatic compounds solutions. The solvent effects on the position and intensity of the $^1A_{1g} \rightarrow ^1B_{2u}$ benzene transitions. *Rev. Roum. Chim.* 20(11–12):1413–1432.

Macovei, V. 1976a. Spectral solvent effects in aromatic compounds solutions. The solvent effects on the position and intensity of chlorobenzene band at 2600 Å. *Rev. Roum. Chim.* 21(2):193–205.

Macovei, V. 1976b. Spectral solvent effects in aromatic compounds solutions: The solvent effects on the position and intensity of nitrobenzene band at 2500 Å. *Rev. Roum. Chim.* 21(8):1137–1147.

McRae, E.G. 1956. Theory of solvent effects on molecular electronic spectra. Frequency shifts. *J. Phys. Chem.* 61:562–572.

Moran, A.M. and Kelleya, A.M. 2001. Solvent effects on ground and excited states structures of p-nitro-aniline. *J. Chem. Phys.* 115(2):912–924.

Murgulescu, I.G. and Sahini, V.E. 1978. *Introducere in Chimia Fizica*. Editura Academiei, Bucuresti.

Nicol, M., Swain, J., Shun, Y.Y., Merin, R., and Chen, R.H.H. 1968. Solvent Stark effect and spectral shifts II. *J. Chem. Phys.* 48(8):3587–3596.

Parasegian, V.A. (ed.) 2006. Van der Waals forces. In *A Handbook for Biologist, Chemists, Engineers, and Physicists*. Cambridge University Press, Cambridge.

Pop, V., Haba, M., and Radu, G. 1978. Einfluss des Losungsmittels auf die electronischen absorptionsspektren der molekule von 4-amino-3-nitro-toluen (4A3NT) orto-nitroanylin (ONA) meta-nitroanylin (MNA) und para-nitroanylin (PNA). *Analele Stiint. Univ. Alexandru Ioan Cuza Iasi, Ib, Fiz.* XXIV:53–60.

Reichardt, C. 2003. *Solvents and Solvent Effects in Organic Chemistry*, 3rd edition. Wiley VCH, Weinheim.

Strat, G. and Strat, M. 2000. Shifts of absorption and fluorescence spectra of anthracene derivatives in binary and ternary solutions. *J. Mol. Liq.* 85:279–290.

Tanakaa, T., Nakajimab, A., Watanabeb, A., Ohnoa, T., and Ozak, Y. 2003. Surface enhanced Raman scattering spectroscopy and density functional theory calculation studies on adsorption o-, m- p-nitroaniline on silver and gold colloid. *J. Mol. Struct.* 661–662:437–449.

2 Solvatochromic Behavior of Ternary Solutions of Some 1,2,4-Triazolium Ylids

Valentina Closca

CONTENTS

Abstract

Triazolium ylids are dipolar molecules in their ground electronic state; as spectrally active molecules, they have a visible absorption intramolecular charge transfer band, very sensitive to the solvent nature. The spectral study of ternary solutions provided information on the specific interaction of 1,2,4-triazolium ylids in protic solvents, the composition of the first solvation shell, and the estimation of the difference between the potential energies in molecular pairs of the types 1,2,4-triazolium ylids—active solvent and 1,2,4-triazolium ylids—inactive solvent. The spectral data achieved with four types of binary solvents assessed on the basis of the statistical model of ternary solutions indicate the nonhomogeneity of the studied ylids ternary solutions: the first solvation shell of the studied triazolium ylids and the rest of the solution has different concentrations of the active solvent molecules.

The results about both intermolecular interaction and the solvent molecule distribution in the vicinity of 1,2,4-triazolium ylids are important due to the fact that triazolium ylids have a wide range of applications in organic chemistry and in the pharmaceutical industry.

2.1 INTRODUCTION

The study of intermolecular interactions in solutions of organic compounds presents a great interest due to the large amount of information that can be obtained: solvent effects, types of interactions, and electro-optical parameters of the spectrally active molecules (dipole moment, polarizability, etc.). The characteristics of molecules can be studied by using various theoretical models and spectral techniques of investigation.

The study of intermolecular interactions presents a wide range of applications in various fields: medicine, chemistry, physics, etc.

The spectral methods used in the investigation of intermolecular interactions in solutions are highly useful, as they allow obtaining information on the internal structure of the liquid, the local fields of forces acting inside the liquids, reaction speed, equilibrium constants, etc.

In spectral research, spectrally active molecules provide information on the intensity of intermolecular interactions by modifying their electronic spectra when passing from gas into liquid. In liquid theory (McRae 1957, Abe 1965, Dorohoi 1994), the spectral modifications are correlated with electro-optical parameters of the solvents, so that, at present, it is possible both to describe the local order in liquid solutions, and to estimate electro-optical parameters of the spectrally active molecule (dipole moment, polarizability).

The behavior of some spectrally active molecules (1,2,4-triazolium ylids) in binary and ternary solutions was studied in view to assess the nature and intensity of intermolecular interactions between them and the liquids they are solved in, as well as to estimate the composition of the first solvation shell of some 1,2,4-triazoli-1-um phenacylids solved in a mixture of two solvents: one active and one inactive from the intermolecular interaction point of view.

The 1,2,4-triazolium ylids can be carbanion mono- or disubstituted. In carbanion, monosubstituted 1,2,4-triazoli-1-um ylids having one hydrogen atom and one phenacyl as carbanion substituents are relatively stable zwitterionic compounds. Triazol-1-ium phenacylids are cycloimmonium ylids in which a hetero-nitrogen is covalently bonded to a carbanion.

The carbanion disubstituted 1,2,4-triazolium ylids are much more stable than the carbanion mono-substituted ones (Zugravescu and Petrovanu 1976). The electronegativity of the substituents increases the stability of triazolium ylids.

1,2,4-Triazolium ylids show a visible absorption band due to an intramolecular charge transfer toward the carbanion to the heterocycle, sensitive to the solvent action and which disappears in acid solutions.

Both universal (long-range) and specific forces (such as hydrogen bond) can act in 1,2,4-triazolium ylid solutions.

Triazolium ylids are used as precursors in obtaining new heterocyclic compounds (Surpateanu et al. 1995, 2000), in pharmacy, as analytical reagents (Melniciuc Puica et al. 2004), as semiconducting (Teles et al. 1996) samples, as substances with antimicrobial and antifungal action (Borowiecki et al. 2013), and so on.

The visible band of triazolium ylids (Surpateanu et al. 1995) is of low intensity, and it is sensitive both to the solvent nature and to the solution pH. This band was used here as an indicator of the solution homogeneity because its position in the wavenumber scale determines the average statistic weight of the solvent in the first solvation shell of the spectrally active molecules.

The difference of the potential energies in molecular pairs of the types active solvent (1)–ylid and inactive solvent (2)–ylid was estimated on the basis of the statistical cell theory of the ternary solutions (Mazurenko 1972, Pop et al. 1986, Avadanei et al. 2011, Dorohoi et al. 2013). Information about the potential energy in the molecular pairs is important in quantum chemistry when the condensed media are described quantitatively. The majority of reactions of ylids are made in solutions; so, knowledge about the intermolecular interactions of these compounds with solvents is very important for developing new applications.

2.2 EXPERIMENTAL

Six carbanion-monosubstituted 1,2,4-triazoli-1-um phenacylids ($TY_1 - TY_6$), having the structures from Scheme 2.1 with substituents from Table 2.1, were prepared from the corresponding salts (Petrovanu et al. 1979, 1983) and purified. The purity of the prepared ylids was checked by elemental analysis and the molecular structures were confirmed by spectral (IR and NMR) analysis.

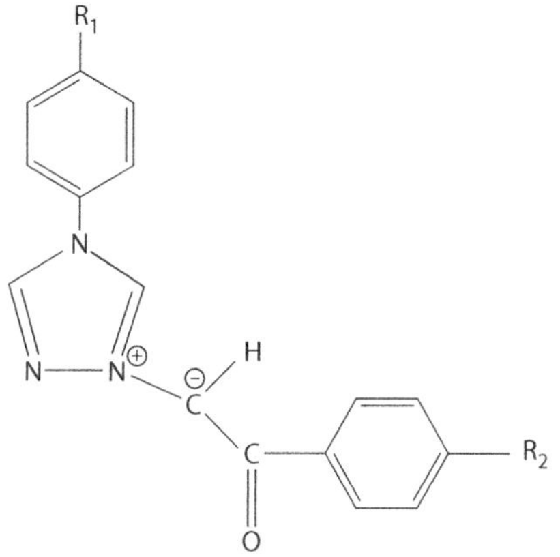

SCHEME 2.1 Chemical structure of the studied 1,2,4-triazoli-1-um ylids.

TABLE 2.1
Substitutes of the Analyzed Structures

Ylid	R_1	R_2	Name
TY_1	–H	–H	4′-Phenyl-1,2,4-triazol-1-ium phenacylid
TY_2	–H	–Cl	4′-Phenyl-1,2,4-triazol-1-ium-(p)-chloro-phenacylid
TY_3	–H	–NO$_2$	4′-Phenyl-1,2,4-triazol-1-ium-(p)-nitro-phenacylid
TY_4	–CH$_3$	–H	4′-Tolyl-1,2,4-triazol-1-ium phenacylid
TY_5	–CH$_3$	–Cl	4′-Tolyl-1,2,4-triazol-1-ium-(p)-chloro-phenacylid
TY_6	–CH$_3$	–NO$_2$	4′-Tolyl-1,2,4-triazol-1-ium-(p)-nitro-phenacylid

The binary solvent was realized in volumetric ratios (C_i, $i = 1,2$) and then the molar concentrations (x_i, $i = 1,2$) were computed, by using the values for density (ρ_i) and molar mass (M) of the two solvents and also the formula (Pop et al. 1986; Dorohoi and Pop, 1987)

$$x_i = \frac{C_i(\rho_i / M_i)}{C_1(\rho_1 / M_1) + C_2(\rho_2 / M_2)}, \quad i = 1,2 \tag{2.1}$$

The ylids were solved directly in the binary solvent in concentrations of about 10^{-4} mol/L (Dorohoi et al. 2008, 2013). In these conditions, the energetic supply of interactions between the spectrally active molecules is very small and can be neglected (Dorohoi 2006; Melniciuc Puica et al. 2014).

The electronic absorption spectra of solutions were recorded on a Specord UV-VIS spectrophotometer with data acquisition system.

2.3   RESULTS AND DISCUSSION

The most stable conformers of carbanion monosubstituted 1,2,4-triazoli-1-um methylids ($TY_1 - TY_6$) were previously established by advanced metering infrastructure (AMI), by searching the minimum of the electronic energy reached in the rotational analysis around the ylid bond (Surpateanu et al. 2003, Dorohoi et al. 2008). The plane of the triazolium ring is perpendicular on the benzene ring directly bonded to it for all studied ylids. The torsion angle of the benzene ring substituted to the carbanion depends on the $- R_1$ structural nature (Melniciuc Puica et al. 2004, Closca et al. 2014a).

Table 2.2 contains information about the wavenumbers in the maximum of the visible absorption band of the $TY_1 - TY_6$ ylids in benzene and in methanol and also about their difference. The dipole moments (Melniciuc Puica et al. 2004) computed by AM1 are also listed in Table 2.2.

From Table 2.2, it results in a linear dependence between the spectral shifts $\Delta \nu = \nu_m - \nu_b$ of the visible absorption band of the studied ylids and their computed dipole moment (Figure 2.1).

TABLE 2.2
Wavenumbers in the Maximum of the ICT Band of the Studied Ylids in Benzene, Methanol, Spectral Shift, $\Delta \bar{\nu} \,(\text{cm}^{-1}) = \bar{\nu}_m - \bar{\nu}_b$, μ (D) Computed by Hyperchem AM1 Dipole Moments of the Studied Molecules

	$\bar{\nu}\,(\text{cm}^{-1})$			
Ylid	Benzene	Methanol	$\Delta \bar{\nu}\,(\text{cm}^{-1})$	μ (D) AMI
TY_1	25,460	25,640	180	7.31
TY_2	25,980	26,360	380	9.06
TY_3	26,110	26,810	700	13.78
TY_4	26,460	26,710	250	10.01
TY_5	26,790	27,120	330	11.01
TY_6	26,910	27,690	780	15.52

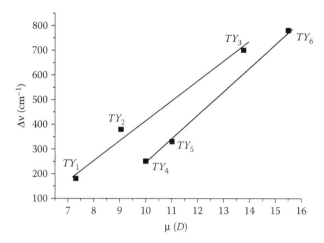

FIGURE 2.1 $\Delta\bar{v}$ (cm^{-1}) versus $\mu(D)$ for the studied triazolium ylids.

TABLE 2.3
Regression Coefficients of the Line Plotted in Figure 2.1

Ylid	Intercept	Slope	R	SD
TY_i (i = 1, 2, 3)	-361.367 ± 103.078	77.748 ± 9.897	0.97	46.843
TY_i (i = 4, 5, 6)	-731.632 ± 37.096	97.288 ± 2.988	0.99	12.406

The ylids placed near each line in Figure 2.1 are characterized by the common heterocycle and the same substituent R_1.

The regression coefficients of the lines from Figure 2.1 are listed in Table 2.3. These dependences suggest the prevalence of the specific and orientation interactions in methanol and indicate an increase of the ylid spectral sensitivity with the molecular dipole moment (Melniciuc Puica et al. 2014).

In order to obtain information about the intermolecular interactions of triazolium ylids with hydroxylic solvents, four ternary solutions of the types active solvent (1) + inactive solvent (2) + ylid were realized and their electronic absorption spectra were recorded. Spectral data were obtained for the ternary solutions:

- 1,2,4-triazolium ylid + methanol (1) + benzene (2)
- 1,2,4-triazolium ylid + water (1) + ethanol (2)
- 1,2,4-triazolium ylid + water (1) + methanol (2)
- 1,2,4-triazolium ylid + 1,3-propane diol (1) + dimethyl formamide (2)

The active solvent from the intermolecular interactions point of view is noted by (1) and the inactive or a little active solvent is noted by (2). The active solvent determines the highest spectral shifts of the visible electronic band of 1,2,4-triazolium

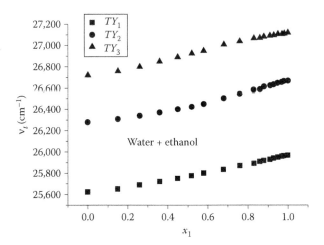

FIGURE 2.2 The wavenumbers in the visible absorption band versus molar fraction of the active solvent in ternary solutions water (1) + ethanol (2) + 1,2,4-triazolium ylid.

ylids binary solutions (Closca et al. 2014a, Melniciuc Puica et al. 2014) relative to benzene solutions.

Modifications of the wavenumbers in the visible electronic absorption bands were observed when the molar concentration of the active solvent (1) is changed by varying its volumetric ratio in ternary solutions (Figure 2.2).

The visible electronic absorption spectra of the studied ternary solutions were recorded and analyzed on the basis of the statistical cell model of the ternary solutions (Closca et al. 2014a,b).

The ternary solution can be imaged as being composed by macroscopic subsystems consisting only from a central (spectrally active) molecule and the first solvation shell around it. The number of the subsystems is proportional with the spectrally active molecule concentration in ternary solution. The thermal motion at room temperature influences, by collisions, the distribution of the solvent molecules around the spectrally active molecules; only the time-averaged composition of the solvation shells will be reflected in the absorption spectrum. In order to compute the potential energy for the interaction between the spectrally active molecule and the solvent molecules from its first solvation shell, one defines the average statistic weight, p_1, of the active solvent in the first solvation shell of the spectrally active molecule (Dorohoi and Pop 1987, Dulcescu et al. 2010, Closca et al. 2014b).

$$p_1 = \frac{\bar{N}_1}{N} = \frac{\nu_t - \nu_2}{\nu_1 - \nu_2} \tag{2.2}$$

In Equation 2.2, $\bar{N}_1$ is the average number of the molecules of the solvent (1) in the first solvation shell, N is the total number of solvent molecules in the first solvation shell, and the wavenumbers ν (cm^{-1}) in the maximum of the visible absorption band of triazolium ylids possess subscript 1, 2 or t corresponding to binary solutions in active solvent (1), in inactive solvent (2), or in ternary solution (t), respectively. The parameter p_1 signifies the average value of the relative number of molecules of the active solvent (1) in the first solvation shell of the spectrally active molecule.

The average statistic weight of the active solvent in the first solvation shell of the solute molecule generally differs from the molar ratio of this solvent in ternary solution. This is because the solvent molecules compete to occupy the favored places near the spectrally active molecules, and so to stabilize the solution. The distribution of the solvent molecules in the first solvation shell of 1,2,4-triazoli-1-um phenacylids is determined by the balance between the intermolecular interactions' energy and the thermal energy.

Weak hydrogen bonds are possible between the −OH groups of the protic solvents and the ylid carbanion possessing a nonparticipant electron pair. Consequently, the binary solvent will act on the complex formed by ylid molecules (Mazurenko 1972, Dulcescu et al. 2010) with hydroxylic solvents.

The nonhomogeneity of the ternary solution, determining its anisotropy, is given by the parameter excess function (Sasirekha et al. 2008)

$$\delta_1 = p_1 - x_1 \qquad (2.3)$$

and preferential solvation constant, k_{12} (Frenkel et al. 1970)

$$k_{12} = \frac{p_1}{x_1} \frac{x_2}{p_2} \qquad (2.4)$$

where p_1 and x_1 are the average statistic weight of the active solvent in the first solvation shell of the spectrally active molecule and the active solvent (1) molar concentration in the binary solvent, respectively.

For positive values of the excess function, the molecules of the active solvent (1) are predominant in the first solvation shell of the central molecule. The negative values of the parameter δ_1 show that the inactive solvent (2) is predominant near the spectrally active molecule. The values of δ_1 differing from zero are indicators of the ternary solution nonhomogeneity.

The same information about the ternary solutions homogeneity is contained in the values of preferential solvation constant, k_{12} (Frenkel et al. 1970) defined by relation (2.4). The active solvent molecules are predominant in the first solvation shell of the spectrally active molecule when the values of k_{12} are higher than unity.

In methanol (1) + benzene (2) + ylid ternary solutions, the ylid complex formed by the hydrogen bond is polar and the methanol molecules possess a minimum of energy when they are placed near it. So, the first solvation shell of the complex methanol–ylid is enriched in methanol molecules compared to the rest of ternary solution. The excess functions δ_1 for ternary solutions methanol (1) + benzene (2) + TY_i ($i = 1$–6) are positive and preferential solvation constants, k_{12} are higher than unity as it results from Figures 2.3 and 2.4.

The relative number of methanol molecules in the first solvation shell of the polar complex formed between the ylid and methanol molecules is higher than in the rest of the solution.

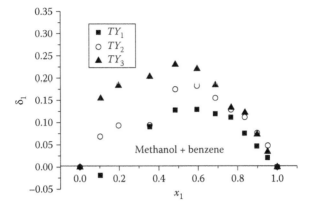

FIGURE 2.3   The excess function δ_1 versus x_1 for ternary solutions methanol + benzene + TY_i ($i = 1,2,3$).

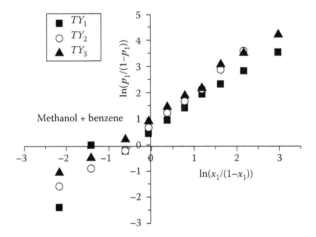

FIGURE 2.4 $\ln p_1/1 - p_1$ versus $\ln x_1/1 - x_1$ for TY_i ($i = 1$–3) in methanol + benzene binary solvent.

The statistic cell model of the ternary (Pop et al. 1986, Dorohoi 2004, Closca et al. 2014b) solutions establishes the linear relation

$$\ln \frac{p_1}{1 - p_1} = m \ln \frac{x_1}{1 - x_1} + n \tag{2.5}$$

in which the slope m is near the unity and the cut at origin n contains information about the difference $w_2 - w_1$ between the potential energies in molecular pairs formed by the two solvents with the ylid molecules.

The linear dependences between $\ln(p_1/1 - p_1)$ and $(x_1/1 - x_1)$ are given in Figure 2.4.

TABLE 2.4

Regression Coefficients for Equation 2.5 for Ternary Solutions Methanol (1) + Benzene (2) + 1,2,4-Triazolium Ylids TY_i, $i = 1$–6

Solvent	Ylid	$m \pm \Delta m$	$n \pm \Delta n$	R	SD	N	$w_2 - w_1 (\cdot 10^{-20} J)$
Methanol + benzene	TY_1	1.054 ± 0.087	0.574 ± 0.139	0.94	0.418	10	0.771 ± 0.187
	TY_2	1.213 ± 0.024	0.801 ± 0.058	0.99	0.171	9	1.076 ± 0.078
	TY_3	1.082 ± 0.033	1.075 ± 0.053	0.99	0.159	10	1.444 ± 0.071
	TY_4	1.289 ± 0.039	1.671 ± 0.048	0.99	0.135	8	2.244 ± 0.064
	TY_5	1.176 ± 0.030	1.515 ± 0.041	0.99	0.120	9	2.035 ± 0.055
	TY_6	1.152 ± 0.032	1.123 ± 0.050	0.99	0.152	10	1.508 ± 0.067

The statistic parameters of lines (2.5) are given in Table 2.4. The cut at the origin gives the difference $w_2 - w_1$ by the relation (Dulcescu et al. 2010, Closca et al. 2014b)

$$n = \frac{w_2 - w_1}{kT} \tag{2.6}$$

In Equation 2.6, k is the Boltzmann constant, T is the absolute temperature, and $w_2 - w_1$ is the difference between the potential energies in molecular pairs of the types methanol–ylid and benzene–ylid.

It results in positive values for the cuts at the origin of lines (2.4) for all studied ylids, indicating $|w_1| > |w_2|$.

The differences $w_2 - w_1$ are listed in the last column of Table 2.4. The values of the difference $w_2 - w_1$ show that the specific interaction between the –OH group of methanol and the ylid carbanion is of the weak hydrogen bond type (Closca et al. 2014b).

In a study of 1,2,4-triazolium ylids intermolecular interactions, the hydroxylic liquids methanol and water (Closca et al. 2014b) were considered, each of the liquids being able to interact specifically with triazolium ylids. If the strength of the specific interactions between the ylid and the two liquids are of the same order of magnitude, this type of interactions will have a small contribution in the difference $w_2 - w_1$.

The values of the excess functions δ_1 (Sasirekha et al. 2008, Closca et al. 2014a,b) and the preferential solvation constants k_{12} (Frenkel et al. 1970, Closca et al. 2014a,b) of water in ternary solutions water (1) + methanol (2) + ylid indicate that the first solvation shell of TY_4 contains a much higher number of water molecules than the rest of the solutions, because the values of δ_1 are positive for all ternary solutions of this ylid

In the ternary solution of TY_5 in the binary solvent water + methanol, for $C_1 < 40\%$, the content in water of the first solvation shell of the ylid molecules is smaller than in the rest of ternary solution ($\delta_1 < 0$). All solvent shells of TY_6 contain a smaller number of water molecules compared to the rest of the solution.

Dependence (2.4) $\ln(p_1/1 - p_1)$ versus $\ln(x_1/1 - x_1)$ is given in Figure 2.5. The regression coefficients computed for lines (2.5) on the basis of the spectral shifts of

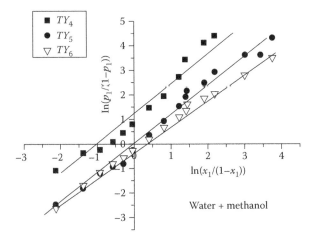

FIGURE 2.5 $\ln p_1/1 - p_1$ versus $\ln x_1/1 - x_1$ for TY_i $(i = 4-6)$ in water + methanol binary solvent.

TABLE 2.5

Regression Coefficients for Equation 2.5 for Ternary Solutions Water (1) + Methanol (2) + 1,2,4-Triazolium Ylids TY_i, $i = 4-6$

Solvent	Ylid	$m \pm \Delta m$	$n \pm \Delta n$	R	SD	N	$w_2 - w_1 (\cdot 10^{-20} J)$
Water + methanol	TY_4	1.355 ± 0.083	1.184 ± 0.109	0.96	0.371	12	0.478 ± 0.044
	TY_5	1.215 ± 0.040	-0.053 ± 0.075	0.98	0.267	16	-0.021 ± 0.030
	TY_6	1.048 ± 0.022	-0.215 ± 0.039	0.99	0.137	15	-0.087 ± 0.016

visible absorption band of ylids in ternary solutions water (1) + methanol (2) + TY_i $(i = 4, 5, 6)$ are listed in Table 2.5 (Closca et al. 2014b).

The difference $w_2 - w_1$ is negative in the cases of TY_5 and TY_6 demonstrating that $|w_2| > |w_1|$. The specific interactions between the ylid and methanol molecules are comparable but they are stronger than the specific interactions between water and ylid (the corresponding ternary solutions give negative cuts at the origin).

The spectral data referring to ternary solutions of the type water (1) + ethanol (2) + TY_i, $i = 1-3$ [24, 27] achieved with two hydroxylic solvents were computed on the basis of the statistical model of ternary solutions.

In the ternary solutions of the type water (1) + ethanol (2) + TY_i, $i = 1, 2, 3$, the excess function δ_1 has negative values for $C_1 < 85\%$ in the case of ylids TY_1 and TY_2 and for $C_1 < 30\%$ in the case of TY_3. It results that in these cases, the molecules of ethanol are prevalent in the first solvation shell of triazolium ylids. For higher volumetric concentrations, the first solvation shell of all triazolium ylids contains a higher number of water molecules compared to the rest of solutions. The values of the preferential solvation constant k_{12} attest the same situations (Table 2.6).

TABLE 2.6
Excess Function δ_1 and Preferential Solvation Constant, k_{12} for Ternary Solutions Water (1) + Ethanol (2) + TY_i (i = 1–3)

	δ_1			k_{12}		
$C_1\%$	TY_1	TY_2	TY_3	TY_1	TY_2	TY_3
0	0.000	0.000	0.000	–	–	–
5	−0.064	−0.073	−0.050	0.531	0.472	0.630
10	−0.074	−0.106	−0.060	0.649	0.517	0.712
15	−0.074	−0.129	−0.035	0.711	0.533	0.856
20	−0.079	−0.142	−0.025	0.722	0.543	0.903
25	−0.077	−0.148	−0.008	0.734	0.546	0.970
30	−0.066	−0.144	−0.005	0.767	0.560	0.980
40	−0.066	−0.103	0.045	0.749	0.642	1.241
50	−0.046	−0.081	0.040	0.789	0.669	1.263
60	−0.059	−0.048	0.045	0.691	0.735	1.434
65	−0.031	−0.065	0.028	0.787	0.631	1.284
70	−0.023	−0.021	0.033	0.818	0.831	1.422
75	−0.024	−0.025	0.028	0.766	0.758	1.484
80	−0.016	−0.020	0.020	0.803	0.763	1.430
85	−0.007	−0.001	0.025	0.868	0.974	2.053
90	0.001	0.004	0.005	1.052	1.175	1.206
95	0.006	0.007	0.008	1.408	1.571	1.612
100	0.000	0.000	0.000	–	–	–

Table 2.6 contains both k_{12} values smaller than unity, showing enrichment in ethanol molecules of the ylid first solvation shell and also values higher than unity ($x_1 > 90\%$ for TY_1 and TY_2 and $x_1 > 40\%$ for TY_3). The prevalence of the water molecules in the ylid first solvation shell is showed by the values of k_{12} higher than unity.

Specific interactions with ylid molecules act in both liquids water and ethanol. Water and ethanol molecules are in competition to occupy the spatial configurations (in the first solvation shell of the spectrally active molecules) corresponding to minimum potential energy of the ternary solution.

In all studied solutions water (1) + ethanol (2) + TY_i, i = 1–6, the ylids are in complex forms of the types water (1)–ylid, or ethanol (2)–ylid. These complexes are made by hydrogen bond interactions between the −OH groups of hydroxylic solvents and the lone electron pair of the ylid carbanion. These complexes interact with the rest of the solution by universal interactions especially of the orientation, induction, or polarization types.

In ternary solutions water (1) + ethanol (2) + TY_i, i = 1–6, the cuts at the origin of line (2.5) are very small and negative for all ylids, because the difference $w_2−w_1$ contains, with opposite signs, the energy of specific interactions in both pairs water–ylid and ethanol–ylid.

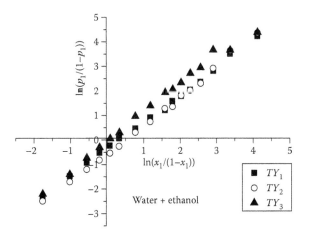

FIGURE 2.6 $\ln p_1 / 1 - p_1$ versus $\ln x_1 / 1 - x_1$ for TY_i ($i = 1$–3) + water (1) + ethanol (2) ternary solutions.

Dependence (2.5) is illustrated in Figure 2.6 for ternary solutions water (1) + ethanol (2) + TY_i, $i = 1$–6, showing a negative cut at the origin for all triazolium ylids under the study.

The regression coefficients of Equations 2.5 in the case of ternary solution water (1) + ethanol (2) + TY_i, $i = 1$–6 are given in Table 2.7. The smallest value of the intercept at the origin was obtained for ylid TY_3 (Table 2.7).

The difference $w_2 - w_1$ is negative for all water (1) + ethanol (2) + TY_i, $i = 1$–3 ternary solutions and it is very small. In conformity with relation (2.3), it results that $|w_2| > |w_1|$. Consequently, it results that the absolute value of the potential energy in pair ethanol (2)–ylid is higher than the potential energy in pair water (1)–ylid.

At low water concentrations in the binary solvent, the complexes of the types ethanol–TY_i, $i = 1$–6 are prevalent, while at high water content, the complexes water–ylid become predominant.

The fourth type of the studied solutions studied was 1,2,4-triazolium ylid + 1,3-propane diol (1) + dimethyl formamide (2); 1,3-propane diol was considered by us as the active solvent, while DMF is a little active solvent for 1,2,4-triazolium ylids. Hydrogen bonds can act in both solvents of this ternary solution, but the most probable is the complex of the type 1,2,4-triazolium ylid–1,3-propane diol.

The wavenumbers in the maximum of the visible absorption band of 1,2,4-triazolium ylid + 1,3-propane diol (1) + dimethyl formamide (2) permitted the study of the composition of the first solvation shell and the nature and the strength of the intermolecular interaction, on the basis of the statistical cell model of the ternary solutions.

The dependences p_1 versus x_1 for binary solvent 1,3-propane diol (1) + dimethyl formamide (2) are illustrated in Figure 2.7 for three 1,2,4-triazolium ylids. From this figure, it results in a saturation of the first solvation shell in 1,3-propane diol molecules for high molar concentrations of 1,3-propane diol in the binary solvent.

TABLE 2.7

Regression Coefficients for Equation 2.5 for Ternary Solutions Water (1) + Ethanol (2) + 1,2,4-Triazolium Ylids TY_i, $i = 1$–6 and for 1,3-Propane Diol (1) + Dimethyl Formamide (2) + 1,2,4-Triazolium Ylids

Solvent	Ylid	$m \pm \Delta m$	$n \pm \Delta n$	R	SD	N	$w_2 - w_1 (\cdot 10^{-20} J)$
Water + ethanol	TY_1	1.000 ± 0.013	-0.369 ± 0.027	0.99	0.086	16	0.119 ± 0.035
	TY_2	1.174 ± 0.017	-0.566 ± 0.034	0.99	0.109	16	-0.229 ± 0.014
	TY_3	1.161 ± 0.026	-0.047 ± 0.052	0.99	0.165	16	-0.019 ± 0.021
	TY_4	1.046 ± 0.020	-0.667 ± 0.040	0.99	0.129	16	-0.270 ± 0.016
	TY_5	1.196 ± 0.018	-0.698 ± 0.037	0.99	0.117	16	-0.282 ± 0.015
	TY_6	1.069 ± 0.039	-0.865 ± 0.079	0.98	0.250	16	-0.350 ± 0.032
1,3-Propanediol + DMF	TY_1	0.506 ± 0.055	-0.131 ± 0.106	0.96	0.123	4	-0.053 ± 0.043
	TY_2	0.564 ± 0.042	-0.238 ± 0.081	0.98	0.094	4	-0.096 ± 0.033
	TY_3	0.725 ± 0.075	-0.345 ± 0.144	0.97	0.167	4	-0.140 ± 0.058
	TY_4	0.587 ± 0.032	-0.227 ± 0.061	0.99	0.071	4	-0.092 ± 0.025
	TY_5	0.525 ± 0.048	-0.264 ± 0.091	0.98	0.106	4	-0.107 ± 0.037
	TY_6	0.609 ± 0.040	-0.321 ± 0.077	0.99	0.090	4	-0.130 ± 0.031

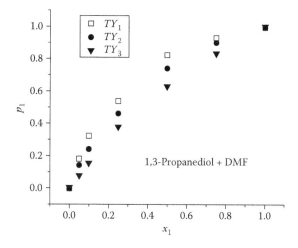

FIGURE 2.7 p_1 versus x_1 for 1,3-propane diol (1) + dimethyl formamide (2) + TY_i ($i = 1, 2, 3$) ternary solutions.

The excess functions δ_1 are all positive, suggesting the prevalence of 1,3-propane diol in the first solvation shell of the studied ylids.

The preferential solvation constants, k_{12}, are higher than unity for all studied ternary solutions 1,3-propane diol (1) + dimethyl formamide (2) + 1,2,4-triazolium ylids, showing that the 1,3-propane diol molecule concentrations are higher in the neighborhood of the ylids, compared to the rest of the ternary solution.

The preferential solvation constants, k_{12}, are higher than unity for all studied ternary solutions 1,3-propane diol (1) + dimethyl formamide (2) + 1,2,4-triazolium ylids, showing that the 1,3-propane diol molecule concentrations are higher in the neighborhood of the ylids, compared to the rest of the ternary solution (Closca et al. 2014a).

For the ternary solutions 1,3-propane diol (1) + dimethyl formamide (2) + TY_i, $i = 1$–6, the cuts at the origin of line (2.5) are negative for all ylids, showing that $|w_2| > |w_1|$. The regression coefficients of the linear dependences $\ln(p_1/1 - p_1) = m$ $\ln(x_1/1 - x_1) + n$ for the 1,3-propane diol (1) + dimethyl formamide (2) + TY_i, $i = 1$–6 solutions are listed in Table 2.7.

The hydrogen bond formation energy is higher in the case of molecular pair 1,2,4-triazolium ylid–DMF, compared to the pair 1,2,4-triazolium ylid–1,3-propane diol (Closca et al. 2014a), as it results from the negative values of the difference $w_2 - w_1$ (see Table 2.7). This fact could be explained by the presence of two hydroxyl groups in diol molecules.

2.4   CONCLUSIONS

The spectral study of ternary solutions of triazolium ylids offers information about the composition of the first solvation shell and the difference between the interaction energies in molecular pairs of the type active solvent–ylid and inactive solvent–ylid. In ternary solutions methanol (1) + benzene (2) + $TY_i = 1$–6, an important part of this difference corresponds to the energy of the hydrogen bond between methanol and ylid molecules, while in ternary solutions achieved with binary hydroxylic solvents, this difference is very small because both the hydroxylic solvents participate in specific and orientation interactions.

The composition of the first solvation shell of the studied ylids was established based on the spectral data obtained for ternary solutions. For all studied ternary solutions of 1,2,4-triazolium ylids, the differences between the molar fractions of the two solvents and the average statistic weights in the first solvation shell of the spectrally active molecule are indicators of the anisotropy of the ternary solution.

REFERENCES

Abe, T. 1965. Theory of solvent effects on molecular electronic spectra. Frequency shifts. *Bull. Chem. Soc. Japan* 30:1314–1318.

Avadanei, M., Dorohoi, D.O., and Melniciuc Puica, N. 2011. Ordering tendency in ternary solutions of pyrydazinium ylids evidenced by electron spectroscopy. *Proceedings of SPIE. 8001 (July)*: http://dx.doi.org/10.1117/12.891518 (art. 80012 K, published).

Borowiecki, P., Milner-Krawczyk, M., and Plenkiewicz, J. 2013. Chemoenzymatic synthesis and biological evaluation of enantiomerically enriched 1-(β-hydroxypropyl) imidazolium- and triazolium-based ionic liquids. *Belstein J. Org. Chem.* 9:516–525. doi: 10.3762/bjoc.9.56.

Closca, V., Melniciuc Puica, N., Benchea, A.C., and Dorohoi, D.O. 2014a. Intermolecular interactions in ternary solutions of some 1,2,4-triazolium ylids studied by spectral means. *Proceedings of SPIE* 9286. art. 92862 T, http://dx.doi.org.//10.1117/12.2063764.

Closca, V., Melniciuc Puica, N., and Dorohoi, D.O. 2014b. Specific interactions in hydroxyl ternary solutions of three carbanion monosubstituted 4'-tolyl-1,2,4 triazol-1-ium-4-R-phenacylids studied by visible electronic absorption spectra. *J. Mol. Liq.* 200:431–438.

Dorohoi, D.O. 1994. *Physics of Liquid State* (in Romanian). Editura Gama, Iasi, Romania.

Dorohoi, D.O. 2004. Electronic spectroscopy of *N*-ylids. *J. Mol. Struct.* 704:31–43.

Dorohoi, D.O. 2006. Electric dipole moments of the spectrally active molecules estimated from the solvent influence on the electronic spectra. *J. Mol. Struct.* 792–793:86–92.

Dorohoi, D.O., Avadanei, M., and Postolache, M. 2008. Characterization of the solvation spheres of some dipolar spectrally active molecules in binary solvents. *Optoel. Adv. Mat. Rapid Comm.* 2:511–515.

Dorohoi, D.O., Dimitriu, D.G., Dimitriu, M., and Closca, V. 2013. Specific interactions in *N*-ylid solutions, studied by nuclear magnetic resonance and electronic absorption spectroscopy. *J. Mol. Struct.* 1044:79–86.

Dorohoi, D.O. and Pop, V. 1987. The spectral shifts in visible electronic absorption spectra of some cycloimmonium ylids in ternary solutions. *An. Stiint. Univ. Al.I. Cuza Iasi 1B Fizica,* Tom XXXIII:78–85.

Dulcescu, M.M., Stan, C., and Dorohoi, D.O. 2010. Spectral study of intermolecular interactions in water ethanol solutions of some carbanion disubstituted pyridinium ylids. *Rev. Roum. Chim.* 55:403–408.

Frenkel, L.S., Laugford, C.H., and Stengle, T.R. 1970. Nuclear magnetic resonance techniques for the study of preferential solvation and the thermodynamics of preferential solvation. *J. Phys. Chem.* 74(6):1376.

Mazurenko, I. 1972. Universal'nie vzaimodeistvii v treh componentov jidkostiah, Optika i Spektroscopia [Universal interactions in liquids with three components]. *Optika i Spektroscopia* T. XXXIII:1060.

McRae, E.G. 1957. Theory of solvent effects on molecular electronic spectra. Frequency shifts. *J. Phys. Chem.* 61(5):562–572.

Melniciuc Puica, N., Barboiu, V., Filoti, S., and Dorohoi, D.O. 2004. Reactivity of some1 (*N*)-[(*para*-R$_2$)-phenacyl]4(*N*)-[(*para*-R$_1$)-phenyl]-1,2,4-triazolium-methylides by UV-VIS, IR, and NMR spectra and molecular modeling. *Spectrosc. Lett.* 37(5):458.

Melniciuc Puica, N., Closca, V., Nechifor, C.D., and Dorohoi, D.O. 2014. Anisotropy of ternary solutions containing 1,2,4-triazoliu-1-um phenacylids studied by spectral means. *Rev. Roum. Chim.* 59(5):359–364.

Petrovanu, M., Luchian, C., Surpateanu, G., and Barboiu, V. 1979. 1,2,4-Triazolium ylures I. Synthèse et stéréochimie des réactions de cycloaddition aux composés à liaison ethylenique active [1,2,4-Triazolium ylids I. Synthesis and stereochemistry of cyclo-addition reaction for compounds with ethylene active bonds]. *Rev. Roum. Chim.* 24(5):733–744.

Petrovanu, M., Luchian, C., Surpateanu, G., Barboiu, V., and Constantinescu, M. 1983. 1,2,4-Triazolium-ylures VIII: Réactions des triazolium phénacylures avec les hétérocu-mulènes [1,2,4 Triazolium ylids VIII. Triazolium phenacylids reactions with hetero-cumulenes]. *Tetrahedron* 39(14):2417–2420.

Pop, V., Delibas, M., and Dorohoi, D.O. 1986. Considerations on the statistic model of the intermolecular interactions in ternary solutions. *An. Stiint. Univ. Al.I. Cuza Iasi 1B Fizica,* Tom XXXII:79–84.

Sasirekha, V., Vanelle, P., and Terme, T. 2008. Solvatochromism and preferential solvation of 1,4-dihydroxy-2,3-dimethyl-9,10-anthraquinone by UV–Vis absorption and laser-induced fluorescence measurements. *Spectrochim. Acta A* 71(3):2006–2009.

Surpateanu, G., Caea, N., Sufletel, L., and Grandcloudon, P. 1995. Synthesis and character-ization of new azatriazolium ylids *Rev. Roum. Chim.* 40:133–136.

Surpateanu, G.G., Vetogen, G., and Surpateanu, G. 2000. A comparative study by AM1, PM3 and ab initio HF/3-21G methods on the structure and reactivity of monosubstituted carbanion 1,2,4-triazolium ylides. *J. Mol. Struct.* 526:143–150.

Surpateanu, G.G., Delatter, F., Woisel, P., Vergoten, G., and Surpateanu, G. 2003. Structure and reactivity of cycloimmonium ylides. Theoretical study in triazolium ylides by AM1, PM3 and DFT procedure methods. *J. Mol. Struct.* 651(1):29–36.

Teles, J.H., Melder, J.P., Gehrer, E., Harder, W., Ebel, K., and Groening, C. Addition products of triazolium salts. U.S. Patent no. 5508422/Apr. 16, 1996.

Zugravescu, I. and Petrovanu, M. 1976. *N-Ylid Chemistry.* Academic Press, McGraw Hill, New York.

3 Spectral Insights on Intermolecular Interactions in Solutions of Some Zwitterionic Compounds

Dana Ortansa Dorohoi
and Dan Gheorghe Dimitriu

CONTENTS

Abstract

The solvatochromic features of some zwitterionic compounds in different solutions allow to establish the nature of intermolecular interactions and to evaluate their contribution to the spectral shifts measured in electronic spectra. Some molecular descriptors of spectrally active molecules were estimated from the solvatochromic analysis.

3.1 INTRODUCTION

The atomic systems (atoms, molecules, molecular complexes, and so on) contain electric charges in a continuous motion. Owing to instantaneous dipole moment created by the fluctuations in the motion of the valence electrons, each atomic system creates an instantaneous radial electric field acting on neighboring systems, even if it is neutral from the electric point of view.

The intensity of the electric field in condensed media depends on the reciprocal actions between the components. Onsager (cited by McRae 1956) introduced the notion of internal reactive electric field in condensed media, suggesting the reciprocal influence between the atomic systems. In the internal reactive electric field, the particles possess potential energy which depends on their chemical nature and on the modifications induced in electro-optical parameters by external agents (pressure, temperature, irradiative fields, etc.).

The internal reactive electric field from liquids is difficult to be measured because any device introduced in liquid can perturb the internal forces. The strength of the reactive internal field in liquids can be evaluated only through the modifications induced in the molecular parameters reflected in electronic spectra.

The molecular spectra offer information regarding the strength of the electric fields acting on the spectrally active molecules introduced in small concentrations in condensed media. Especially the electronic (absorption and fluorescence) spectra based on modifications induced by light in the valence molecular electronic cloud offer global information regarding the reactive electric field action on the liquid components. In this aim, spectrally active molecules are introduced in small amounts (10^{-3}–10^{-5} mol/L) as "probes" able to give information about the local reactive electric field in the studied liquid. The liquid must solve the spectrally active molecules and must be transparent in the spectral range of the solute spectra.

The spectral changes induced by liquid are related to the band position (given by the wavenumber $\bar{\nu}$ (cm^{-1}) in the band maximum) and/or to the spectral intensity of the electronic band.

3.2 INTERMOLECULAR INTERACTIONS IN LIQUID SOLUTIONS

The intermolecular interactions from liquids can be classified (McRae 1956, Mataga and Kubota 1970, Bakhshiev 1972, Suppan 1990) into universal (long range) and specific (short range, or quasi-chemical) interactions.

The first type of interactions exerts a global influence on the spectrally active molecules. They are nonoriented, nonsaturated, and can cause spectral shifts of the electronic (absorption and/or fluorescence) bands.

The specific interactions consist of processes with charge transfer between the atomic systems such as hydrogen bonds, or electron transfer. Specific interactions act locally, being evidenced especially in vibration (IR or Raman) spectra, but they can cause supplementary spectral shifts in electronic spectra directly affected by universal forces.

Many chemical and physical processes take place in solutions, being influenced by the solvent nature. The solvent effects depend on the chemical structure of the solute molecules and on the magnitude of the changes in their electro-optical parameters (electric dipole moment, polarizability, ionization potential, etc.) induced by external agents, such as the irradiative fields.

The knowledge of the nature and strength of intermolecular interactions in liquids is very important because most of the chemical reactions take place "*in situ*" being directly affected by the solvent nature (Wypych 2001, Reichardt 2003).

Usually, the solutions for which the electronic spectra are recorded contain small amounts of spectrally active molecules. In these conditions, the distances between the spectrally active molecules are too large that the interactions between them are negligible.

The solvent molecules create concentric shells having the solute molecule in their center. The internal electric field acting in the point in which the spectrally active molecule is placed modifies the charge distribution and electro-optical parameters of this molecule. The relative distances between the energetic levels of the solute molecule are also modified in the solvation process.

At room temperature, the molecules composing a studied solution are in their ground state. In the light interaction process, the solvent keeps its ground state (being transparent in the searched spectral range), while the solute molecules absorb radiation, passing in excited electronic states. According to Frank–Condon principle (Bayliss and McRae 1954), the electronic transitions are very quick. So, after absorption, the excited solute molecule is in a destabilized state because the solvent arrangement corresponds to the energetic conditions of its ground state. The solvent relaxation process needs a time interval higher compared to the life time of the excited state. In the case of the excited states with a long life time, the solvent molecules can reorient in the electric field created by the solute-excited state dipole moment and the solute molecule becomes stabilized by the new arrangement of the solvent molecules. This phenomenon modifies the energetic levels of the spectrally active molecules and changes the wavenumber in the maximum of the electronic absorption band.

The absorption process takes place between two electronic levels of the spectrally active molecule. The stabilization energies are usually different in the two electronic levels responsible for the absorption band appearance. Their difference determines the spectral shift of the electronic band in the wavenumber scale relative to its position in the gaseous phase corresponding to the isolated molecule (Bakhshiev 1961, Mataga and Kubota 1970, Stone 2013).

The spectral method offering information on intermolecular interactions in solutions is noninvasive; it does not change the internal equilibrium of forces because spectrally active molecules are used as "probes," no big devices are compared to the molecular dimensions.

In the first stage of the research on intermolecular interactions in solutions, multiple attempts were made in order to evidence some correlations between the spectral characteristics and the micro- and macroscopic parameters of solutions. The spectral shifts were correlated with the solvent refractive index (n), electric permittivity (ε), ionization potential (I), and so on, in order to establish the general characteristics of intermolecular interactions in liquid solutions.

Unfortunately, the liquid phase is very complex, both by the chemical structure of its components and by gentle equilibrium between the potential energy of the molecules in the internal reactive field and the energy of thermal motion which continuously modifies the composition of the molecular shells surrounding the spectrally active molecules and so, the solvation energy of each solute molecule. This fact can be easily evidenced in multicomponent solutions.

The liquid complexity is the cause that the developed theories regarding intermolecular interactions cannot be applied on a large category of liquids.

Some models (McRae 1956, Bakhshiev 1961, Abe 1965) developed for various liquids have limited applicability due to some simplifying hypotheses corresponding to the type of liquid for which they are made. For example, in most of the theoretical descriptions, the specific interactions are neglected, and even there are experimental evidences for their presence in the majority of liquids (Reichardt 2003).

In order to accord the theory with the experiment, the researchers added empirical terms to the theoretical relations to describe the influence of the specific interactions on the electronic spectra.

Some known empirical scales (Kosower 1958, Kamlet et al. 1977, Catalan 2009) were proposed. The solvents are arranged according to their spectral influence on the related compounds from the structural point of view (Benchea et al. 2014, Babusca and Dorohoi 2016).

3.3   COMPUTATIONAL METHODS IN DESCRIBING INTERMOLECULAR INTERACTIONS IN SOLUTIONS

In the specialized literature, there are tendencies to select only the results that confirm the considered models and in which the supplementary spectral shifts are attributed to various specific interactions. This procedure makes such verifications inconvenient and could be considered as nonobjective.

Based on the theoretical and empirical dependences between the spectral shifts and the solution parameters, computational methods were applied to experimental data in order to establish the contribution of different types of interactions to the wavenumber in the maximum of the electronic (absorption and/or fluorescence) bands. Relations of the type

$$\bar{\nu} = \bar{\nu}_0 + C_1 f(\varepsilon) + C_2 f(n) + C_3 \beta + C_4 \alpha + \cdots \tag{3.1}$$

are usually applied to experimental data in order to establish the regression coefficients $C_1, C_2, \ldots, C_n$, which in existent theories are dependent on the microscopic parameters of the spectrally active molecules (Dorohoi 2010).

The regression parameters show, by their magnitude and sign, the sensibility of the spectrally active molecule to the interactions described by the corresponding term and the sense of the spectral shift in the wavenumber scale, respectively.

The greatest value of one regression coefficient demonstrates that the intermolecular interaction described by the corresponding term is dominant in solution.

The negative signs of the terms in relation (3.1) determine the bathochromic (toward red) shift of the electronic bands due to the corresponding effects. Contrarily, the positive signs of the terms in Equation 3.1 determine the hypsochromic effect, shifting the electronic bands to blue.

Great correlation coefficient (R) and small standard deviation (SD), resulted from the statistical analysis, demonstrate the applicability of the model (described by the non-null regression coefficients C_i, $i = 1, 2, ..., n$) of Equation 3.1 to the studied solution.

3.4 ZWITTERIONIC COMPOUNDS

The multitude of compounds in organic chemistry is essentially based on the covalent bonds among atoms. The covalent bonds between atoms with different chemical structure have polar character, due to the higher electron density near the more electronegative atom. The chemical compounds containing basic and acid groups able to neutralize each other are named zwitterionic. The structure of a zwitterionic compound can be written with one negative and one positive charges placed on different atoms.

3.4.1 CYCLOIMMONIUM YLIDS AS ZWITTERIONIC COMPOUNDS

The ylidic bond is realized between two charged (one heteroatom and one carbanion) atoms having an orbital occupied by two electrons uninvolved in the covalent bond. The term ylid (Johnson 1966, Zugravescu and Petrovanu 1976) was introduced by Witting in 1944. It suggests both the presence of a free valence and the anionic nature of this kind of molecules.

N-ylids are compounds in which a nitrogen atom is covalently bonded to carbanion. If the nitrogen atom belongs to a heterocycle (see Figure 3.1a), the compounds are named cycloimmonium ylids (Zugravescu and Petrovanu 1976).

As cation, the positive part of the cycloimmonium ylids (Johnson 1966, Zugravescu and Petrovanu 1976) can be one heterocycle, most common being pyridine, iso-quinoline, pyridazine, phthalazinium, and so on.

Having in view the carbanion structure, cycloimmonium ylids can be classified into carbanion mono- and carbanion disubstituted ones, if they possess one proton or one atomic group as R_1 substituent, respectively. The R_2 substituent must be a high electronegative atomic group in order to assure the ylid stability. The carbanion

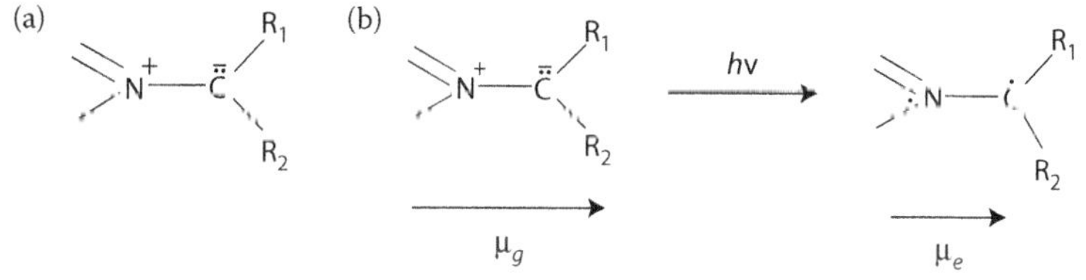

FIGURE 3.1 (a) Chemical structure of cycloimmonium ylids; (b) ICT mechanism in the case of cycloimmonium ylids.

symmetrically disubstituted cycloimmonium ylids has two identical substituents $R_1 = R_2$. Symbiosis effect has been emphasized in the case of the symmetrically substituted carbanion (Surpateanu and Dorohoi 1977, Dorohoi, 2007).

The cycloimmonium ylid stability depends on the chemical nature of both the heterocycle and the groups covalently bonded to the ylid carbanion. The greater the capacity of the groups attached to the carbanion to accept the negative charge and of the heterocycle to bear positive charge, the higher the ylid stability is. The carbanion-disubstituted cycloimmonium ylids are more stable than the carbanion-monosubstituted ones. The highest stability is obtained for the carbanion disubstituted with strongly electronegative atomic groups (Surpateanu et al. 1975, Dorohoi 2004).

Cycloimmonium ylids are nucleophilic compounds (Gheorghies et al. 2008, Dorohoi et al. 2013), which can react with a large variety of organic compounds. As a participant to cycloaddition reactions, the cycloimmonium ylids are precursors (Melnig et al. 2006) for new heterocycle compounds.

3.4.2 SPECTRAL STUDY OF CYCLOIMMONIUM YLIDS USING THE SOLVENT EMPIRICAL POLARITY

In absorption, all cycloimmonium ylids show a visible electronic band classified into an intramolecular charge transfer (ICT) (see Figure 3.1b) from the carbanion toward the heterocycle (Dorohoi et al. 1994, 2012, Dimitriu et al. 2008, Closca et al. 2014a). This electronic band is very sensitive to the solvent nature and it disappears in acid media, due to ylid protonation at the carbanion level (Gheorghies et al. 2008, Dorohoi et al. 2013).

In chemistry, the cycloimmonium ylids can be used as acid–basic indicators because this band is very sensitive to the solution pH (Zugravescu and Petrovanu 1976, Dulcescu and Dorohoi 2009).

The linear dependence of E_{max} (kcal/mol) versus Z (kcal/mol) (Figure 3.2 and Tables 3.1 and 3.2) proves the ICT nature of the visible band of cycloimmonium ylids (Dorohoi 2004, Dulcescu et al. 2010a,b, Homocianu et al. 2011a,b). This assertion is based on the ICT in 1-ethyl-4-carbomethoxy-pyridinium iodide responsible for the visible band considered (Kosower 1958, 1968) as an indicator of the solvent polarity.

Being very sensitive to the solvent nature, the visible ICT band of cycloimmonium ylids can give information about the nature and the strength of intermolecular interactions in solutions (Gheorghies et al. 2010, Closca et al. 2014a). So, the cycloimmonium ylids can be used as sounders in estimating the internal reactive field created by the solvent in the place of the solved molecule (Figure 3.3).

The slopes and the cuts at the origin of the equations

$$E_{max} \ (kcal/mol) = mZ \ (kcal/mol) + n \tag{3.2}$$

depend both on the heterocycle nature and on the carbanion substituents (Dorohoi 2004, Closca et al. 2014b, Babusca and Dorohoi 2016). The carbanion mono-substituted cycloimmonium ylids are a few sensitive to the solvent nature (see Table 3.2).

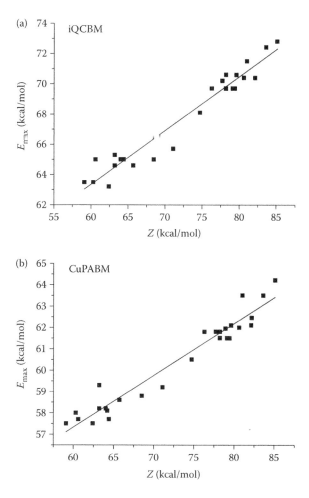

FIGURE 3.2 E_{max} (kcal/mol) versus Z (kcal/mol) for iQCBM (a) and CuPABM (b) in binary solutions.

3.5   SPECTRAL SHIFTS IN CYCLOIMMONIUM SOLUTIONS DESCRIBED BY THEORIES OF LIQUID SOLUTIONS

As dipolar molecules, cycloimmonium ylids participate essentially to orientation–induction interactions which are stronger in their ground state, compared to the excited electronic state. This fact can be explained due to the charge transfer from carbanion to the heterocycle, causing the decrease of the ylid dipole moment. The cycloimmonium ylids are also nucleophilic compounds able to participate to proton changes with hydroxyl solvents, accepting protons from acidic (HBD) solvents such as alcohols or acids. This phenomenon was demonstrated by the aspects of the graphs of the wavenumber $\bar{v}$ (cm^{-1}) in the maximum of the ICT band versus electric permittivity of the solvents (see Figure 3.4). In these graphs (Dorohoi 2004, 2006,

TABLE 3.1
Energy in the Maximum of the Visible ICT Band of Some Cycloimmonium Ylids in Binary Solutions

Solvent	Z (kcal/mol)	ε	iQCM	iQDiCM	iQCBM	CuPNiPY	CuPABM
			\multicolumn{5}{c}{F_{max} (kcal/mol)}
Chlorobenzene	59.1	5.62	51.0	60.8	63.5	57.0	57.5
Ethyl acetate	60.3	6.02	–	61.5	63.5	56.9	58.0
Anisol	60.6	4.33	–	62.1	65.0	56.9	57.7
Benzene	62.4	2.27	51.5	60.5	63.2	–	57.5
Chloroform	63.2	4.81	51.7	63.0	65.3	57.6	59.3
Dichloroethane	63.2	10.36	–	62.5	64.6	–	58.2
Pyridine	64.0	12.4	51.5	61.5	65.0	56.9	58.2
Dichloromethane	64.2	8.93	51.7	62.8	65.0	57.7	58.1
Iso-amyl acetate	64.4	4.8	–	62.1	65.0	57.8	57.7
Acetone	65.7	20.7	–	62.8	64.6	58.0	58.6
Dimethylformamide	68.5	36.7	51.6	63.5	65.0	57.6	58.8
Dimethylsulfoxide	71.1	46.7	51.4	64.2	65.7	57.8	59.2
Diacetonalcohol	74.7	18.2	51.9	65.7	68.1	58.9	60.5
Iso-propyl alcohol	76.3	17.9	–	65.7	60.7	59.4	61.8
n-Butyl alcohol	77.7	17.51	51.3	66.5	70.2	60.2	61.8
n-Hexyl alcohol	78.2	12.5	51.3	65.7	69.7	59.2	61.5
n-Propyl alcohol	78.2	20.33	–	66.5	70.6	59.6	61.8
n-Octyl alcohol	79.1	9.80	51.3	65.7	69.7	59.2	61.5
n-Amyl alcohol	79.4	14.8	51.8	65.7	69.7	59.6	61.5
Ethanol	79.6	24.5	51.3	67.7	70.6	60.0	62.1
Iso-butyl alcohol	80.6	18.1	51.4	66.5	70.4	59.6	62.0
Formamide	81.0	84.00	52.0	68.9	71.5	59.6	63.5
n-Benzyl alcohol	82.1	13.5	51.6	66.9	70.4	–	62.1
Methanol	83.6	32.7	51.5	68.5	72.4	–	63.5
Ethylene glycol	85.1	37.70	51.7	68.9	72.8	60.6	64.2

Significance of the notations: iQCM, iso-quinolinium carbethoxy methylid; iQDiCM, iso-quinolinium dicarbethoxy methylid; iQCBM, iso-quinolinium carboethoxy-benzoyl-methylid; CuPNiPY, p-cumyl-pyridazinium-p-nitrophenacylid; CuPABM, p-cumyl-pyridazinium-acetyl-benzoyl-methylid.

Dimitriu et al. 2008, Gheorghies et al. 2008, Homocianu et al. 2011a,b) for the same value of the electric permittivity, the protic solvents are placed to higher wavenumbers on the second curve expressing this dependence. At the same time, the aprotic solvents determine the first curve placed on the smaller values of the wavenumbers (see Figure 3.4a and b).

The distance between the two curves containing the aprotic and protic solvents, respectively, can be considered as a measure for the specific interactions of the hydrogen bond type between the ylid and hydroxyl solvent. This distance is proportional

TABLE 3.2

Energy in the Maximum of the Visible ICT Band of Some Cycloimmonium Ylids in Ternary Solutions Ethanol + Water and Acetone + Water

	Z (kcal/mol)	E_{max} (kcal/mol) CuPNiPY	E_{max} (kcal/mol) CuPABM
C% Ethanol in Water			
98	80.2	59.7	62.5
96.9	80.8	60.0	62.7
95.0	81.2	60.1	62.8
92.0	82.0	60.2	62.9
90.0	82.5	60.2	63.0
85.0	83.8	60.3	63.5
80.0	84.8	60.4	63.8
75.0	85.7	60.6	64.2
70.0	86.4	60.7	64.4
C% Acetone in Water			
99.0	68.1	58.3	58.9
95.0	72.9	58.8	59.6
93.0	74.8	58.9	60.0
90.0	76.6	59.1	60.5
85.0	78.8	59.7	61.0
75.0	80.7	60.0	61.8
70.0	82.1	60.3	62.5
65.0	83.2	60.6	62.5
60.0	84.3	60.9	63.2
55.0	85.5	61.2	63.5

with the contribution of the specific interactions to the total spectral shift recorded in the electronic absorption spectra of cycloimmonium ylids.

3.6 SOME DESCRIPTORS OF CYCLOIMMONIUM YLIDS ESTIMATED FROM SOLVATOCHROMIC STUDY

The solvatochromic studies corroborated with the theories about the simple liquids can give information about the strength of the molecular interactions in liquid solutions. The electro-optical parameters in the excited states of the spectrally active molecules can be estimated by the magnitude of the frequency shifts in electronic spectra. The existent theories regarding the solvent influence on the electronic absorption and fluorescence spectra permit to estimate some electro-optical parameters in the molecular-excited states. The theory developed by McRae and modified by Bakhshiev permits to estimate the dipole moments of the spectrally active (in

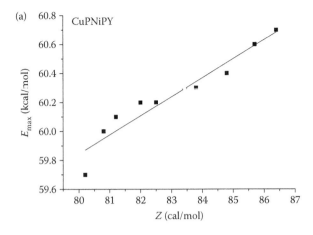

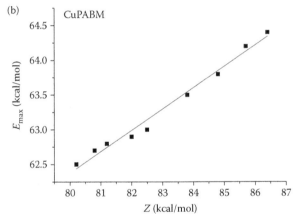

FIGURE 3.3 E_{max} (kcal/mol) versus Z (kcal/mol) for CuPNiPY and CuPABM in ternary solutions ethanol + water + cycloimmonium ylids.

absorption and/or fluorescence) molecules in both electronic states participating to the electronic transitions. In relation (3.1), the significance of the regression coefficient C_1 is given by relations

$$C_1^a = \frac{2}{hcr^3} \frac{2n^2+1}{n^2+2} \mu_g (\mu_g - \mu_e \cos\phi) \qquad (3.3)$$

$$C_1^f = \frac{2}{hcr^3} \frac{2n^2+1}{n^2+2} \mu_e (\mu_e - \mu_g \cos\phi) \qquad (3.4)$$

In relations (3.3) and (3.4), the parameters have the following significance: μ is the electric dipole moment in the ground (g) and in excited (e) electronic states of the solute molecule; ϕ is the angle between the dipole moments in the electronic states participating to the absorption/fluorescence transition, h is Planck's constant, c is the velocity of light, and r is the solute Onsager radius.

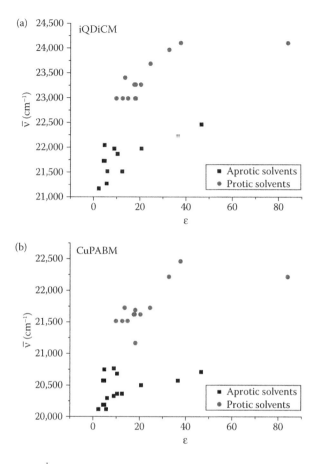

FIGURE 3.4 $\bar{v}$ (cm^{-1}) versus ε for iQDiCM and CuPABM cycloimmonium ylids in binary solutions.

For the molecules with electronic absorption (a) and fluorescence (f) bands, such as the pyridazinium ylids, both the ground state and the excited state dipole moments can be estimated, as it is described below (Dorohoi 2006, Dorohoi et al. 1998). The studied molecules have the structural formula from Figure 3.5 and the attached substituents from Table 3.3. Pyridazinium ylids from Table 3.3 show visible absorption and fluorescence bands (Pop et al. 1994).

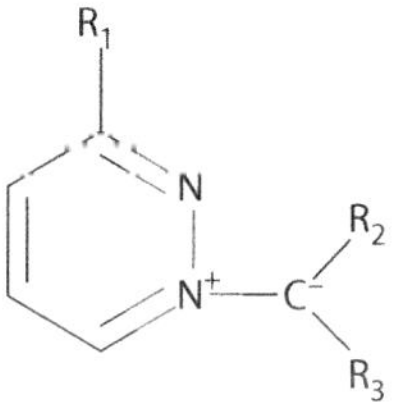

FIGURE 3.5 Chemical structures of the fluorescent pyridazinium ylids.

TABLE 3.3
Pyridazinium Ylids under the Study

Name	R_1	R_2	R_3
(PMBP)	$-H$	H	$-COC_6H_4-OCH_3\ (p)$
(PPMBP)	$-C_6H_5$	$-H$	$-COC_6H_4-OCH_3\ (p)$
(TPDiBM)	$-C_6H_4-CH_3(p)$	$-COC_6H_5$	$-COC_6H_5$
(CPABM)	$-C_6H_4-CH(CH_3)_2(p)$	$-COCH_3$	$-COC_6H_5$

The cycloimmonium ylids' denomination from Table 3.3: Pyridazinium-(p-methoxy-benzoyl) phenacylid (PMBP); p-phenyl-pyridazinium-(p-methoxy-benzoyl)-phenacylid (PPMBP); p-tolyl-pyridazinium-di-benzoyl methylid (TPDiBM); and p-cumyl-pyridazinium-(acetyl-benzoyl) methylid (CPABM).

The wavenumber in the maximum of the ylids from Table 3.3 was correlated with the solvent electric permittivity (ε), refractive index (n), and spectral shift of the hydroxyl proton measured in nuclear magnetic resonance (NMR) spectra. The dependence of the type

$$\bar{v} = \bar{v}_0 + C_1 \frac{\varepsilon - 1}{\varepsilon + 2} + C_2 \frac{n^2 - 1}{n^2 + 2} + C_3 \delta_{OH} \tag{3.5}$$

where δ_{OH} is the chemical shift (in ppm) measured in the NMR spectrum of the protic solvent.

The statistical analysis using relations (3.3) and (3.4) both in absorption and fluorescence emphasized the regression coefficients from Table 3.4.

Table 3.5 contains the values of the electric dipole moment in the ground state and the molecular radius of pyridazinium ylids (computed by HyperChem 8.0.6 using PM3). The excited state dipole moments, the angle φ between dipole moments, and the transition dipole moment are also listed in Table 3.5.

From the data of Table 3.5, it results in the decrease of the molecular dipole moment in the absorption process. The small values of the dipole moment in the excited state determine the dispersion–polarization nature of the intermolecular forces between the solute and the solvent molecules, while in the ground state of pyridazinium ylids, the dipolar orientation–induction interactions are dominant. The solvation energy by orientation–induction interactions is smaller in the ylid-excited state than in its ground state. For similar structures, the same conclusions can be drawn (Pop et al. 1994, Gheorghies and Dorohoi 2008).

The reciprocal orientation of the dipole moments in the electronic states participating in the photon absorption is shown in Figure 3.6.

In order to estimate the dipole moments in the electronic states participating in the absorption/emission process responsible for the visible band appearance, Equations 3.3 and 3.4 were solved having in view the values of the coefficients in multilinear regression applied to the spectral data. In accordance with Table 3.5 and

TABLE 3.4
Regression Coefficients, Correlation Coefficient, and Standard Deviation for Spectral Absorption and Fluorescence Data of the Studied Pyridazinium Ylids Analyzed with Formula (3.1)

Ylid	PMBP	PPMBP	TPDiBM	CPABM
$\bar{\nu}_0^a \pm \Delta\bar{\nu}_0^a$ (cm^{-1})	20550 + 280	19870 + 130	20200 + 150	20080 + 130
$\bar{\nu}_0^f \pm \Delta\bar{\nu}_0^f$ (cm^{-1})	19710 ± 360	19310 ± 260	19560 ± 260	19666 ± 360
$C_1^a \pm \Delta C_1^a$ (cm^{-1})	867 ± 330	683 ± 351	829 ± 208	720 ± 280
$C_1^f \pm \Delta C_1^f$ (cm^{-1})	0	0	0	0
$C_2^a \pm \Delta C_2^a$ (cm^{-1})	9±6	0	0	0
$C_2^f \pm \Delta C_2^f$ (cm^{-1})	−1076 ± 520	−1920 ± 700	−2016 ± 750	−5060 ± 2318
$C_3^a \pm \Delta C_3^a$ (cm^{-1})	0	0	176 ± 27	192 ± 16
$C_3^f \pm \Delta C_3^f$ (cm^{-1})	0	37 ± 12	0	42 ± 22
R^a	0.93	0.95	0.91	0.99
R^f	0.91	0.90	0.93	0.92
SD^a	76	80	96	130
SD^f	88	105	77	117

TABLE 3.5
Ground μ_g (D) and Excited μ_e (D) States Dipole Moments, Angle φ (degree) between Them, and the Transition Dipole Moment $\Delta\mu_{eg}$ (D) of the Studied Pyridazinium Ylids

Ylid	PMBP	PPMBP	TPDiBM	CPABM
μ_g (D) (computed by PM3)	8.30	5.20	8.10	6.50
a (Å) (computed by PM3)	10	8	10	9
μ_e (D)	1.78	1.57	1.66	0.73
φ (degree)	77.62	76.04	78.15	81.90
$\Delta\mu_{eg}$ (D)	7.73	6.30	7.58	5.14
γ (degree)	167.62	166.15	168.15	171.56
μ_g (D) for $\mu_e = 0$	8.11	5.15	7.93	6.31

Equations 3.3 and 3.4, there are two solutions for the excited state dipole moment; the first corresponds to the null value and the latter to $\mu_e = \mu_g \cos\phi$. The values of the ground-state dipole moment in the case $\mu_e = 0$, obtained for the system of Equations 3.3 and 3.4, are written in the last row of Table 3.5.

The hydrogen bonds possible to be made between the nonparticipant electrons of the ylid carbanion and the hydroxyl groups of alcohol or acids are weaker in the excited state of the cycloimmonium ylid molecules. The estimation of a null contribution of the orientation–induction forces in fluorescence spectra proves that by

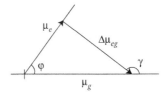

FIGURE 3.6 Reciprocal orientations of the dipole moments in the ground and excited states responsible for the ylid ICT visible band appearance. $\mu_g(D)$ and $\mu_e(D)$ are the dipole moments in the ground and excited states participating to transition, φ(degree) is the angle between them and $\Delta\mu_{eg}(D)$ is the dipole moment of the transition.

excitation both in protic and in aprotic solvents, the electric dipole moment of cyclo-immonium ylids is smaller than the dipole moment in their ground state.

3.7 TERNARY SOLUTIONS OF CYCLOIMMONIUM YLIDS INTERACTION ENERGY IN THE GROUND STATE OF THE MOLECULAR PAIRS

Some interesting results can be obtained in solvatochromic analysis of the ternary solutions of cycloimmonium ylids. In this kind of studies, the difference between interactions energies in the molecular pair made by the solute molecule and the molecules of the two miscible solvents (one active and one inactive from the molecular interactions point of view) is estimated. One can exemplify by the ternary solutions made in a binary solvent containing one protic and one aprotic liquid. As nucleophilic molecules cycloimmonium ylids participate to specific interactions of the type of hydrogen bonds and in this context, the protic solvent can be considered more active than the aprotic ones from the interactions point of view.

As it was established in Chapter 2, the cell model of the ternary solutions (Mazurenko 1972, Perov 1980, Pop et al. 1986) can be applied to some cycloimmonium ylids having triazolium derivatives as the heterocycle. The similar dependences with those established in Chapter 2 can be obtained for other cycloimmonium ylids. We resume only the values of the potential energy in molecular pairs of the types cycloimmonium ylid–active solvent and cycloimmonium ylid–inactive solvent. Estimations (see Tables 3.6 through 3.8) are made for the experimental results obtained in diluted solutions and refer to the ground state of the molecular pairs, as it results from the theoretical model (Pop et al. 1986, Dorohoi and Pop 1987).

When one protic and one aprotic solvents compose the binary solvent, the difference between the molecular pairs energy is proportional to the energy of the hydrogen bond, especially when the macroscopic parameters of the two solvents have appropriate values (in order to equalize the universal interactions of the solute molecule with the two solvents) (Dorohoi et al. 2008, Avadanei et al. 2011, 2015).

In the case of a binary solvent containing two hydroxyl components, the difference $\omega_2 - \omega_1$ is very small, because the specific interactions have contributions in the two terms of the differences (Dulcescu et al. 2010a,b).

TABLE 3.6
Difference of Energies in Molecular Pairs Realized by Phthalazinium Dibenzoyl Methylid (Melniciuc-Puica et al. 2012) and Solvent Molecules in Ternary Solutions

Binary Solvent	$m \pm \Delta m$	$n \pm \Delta n$	R	$kT(n \pm \Delta n)10^{-20}$ (J)
Water + ethanol	0.674 ± 0.049	0.605 ± 0.188	0.998	0.245 ± 0.076
Propionic acid + chloroform	1.8583 ± 0.096	3.587 ± 0.178	0.997	1.451 ± 0.072
Octanol + dichloroethane	1.267 ± 0.010	3.055 ± 0.022	0.999	1.236 ± 0.009

TABLE 3.7
Difference of Energies in Molecular Pairs Realized in Ternary Solutions by Cycloimmonium Di-Carbethoxy Methylids (Having Pyridinium and Iso-Quinolinium as Heterocycles)

Ylid	$m \pm \Delta m$	$n \pm \Delta n$	R	$kT(n \pm \Delta n)10^{-20}$ (J)
PDCM	0.976 ± 0.030	-0.093 ± 0.046	0.99	-0.037 ± 0.018
iQDCM	0.953 ± 0.021	-0.285 ± 0.032	0.99	-0.115 ± 0.013

Source: Dorohoi et al. 2013. *J. Mol. Struct.* 1044:79–86.
PDCM—pyridinium dicarboethoxy-methylid. iQDCM—iso-quinolinium-dicarbo-ethoxy-methylid.
Binary solvent: ethylene glycol + dioxane.

TABLE 3.8
Difference of Energies in Molecular Pairs Realized in Ternary Solutions by Cycloimmonium Carboethoxy-Anilido-Methylids (Having Pyridinium and Iso-Quinolinium as Heterocycles)

Ylid	$m \pm \Delta m$	$n \pm \Delta n$	R	$kT(n \pm \Delta n)10^{-20}$ (J)
PCAM	1.229 ± 0.039	3.326 ± 0.089	0.99	1.345 ± 0.036
iQCAM	1.325 ± 0.078	3.714 ± 0.178	0.98	1.503 ± 0.071

Source: Adapted from Closca, V., Ivan, L.M., and Dorohoi, D.O. 2014a. *Spectrochim. Acta A: Mol. Biomol. Spectrosc.* 122:670–675.
PCAM—pyridinium carboethoxy anilido methylid. iQCAM—iso-quinolinium carbo-ethoxy anilido methylid.
Binary solvent: octanol + dichloromethane.

In Chapter 2, some results obtained for the carbanion monosubstituted triazolium ylids in ternary solutions are detailed.

3.8 CONCLUSIONS

Zwitterionic compounds show visible electronic bands with ICT. They are very sensitive to the solvent nature. The energies in the maximum of the ICT bands of cycloimmonium ylids linearly depend on the solvent empirical polarities defined by Kosower.

The protic and aprotic solvents are separated on two distinct curves in the graphs $\bar{v}$ (cm^{-1}) versus ε due to the supplementary shift caused by hydrogen bonds between the cycloimmonium ylids carbanion and the hydroxyl group of the protic solvents.

When cycloimmonium ylids are spectrally active both in absorption and fluorescence spectra, the dipole moments in the electronic states participating to the electronic transition and the angle between them can be estimated from the solvent influence of the solvent on electronic spectra.

Spectral study in ternary solutions of cycloimmonium ylids permits to estimate, in the limits of the cell model of multicomponent solutions, the difference between the energies in molecular pairs achieved between ylid molecules and the molecules of the two components of the binary solvent. If the binary solvent contains one protic and one aprotic solvent, the energy difference is proportional to the contribution of the hydrogen bond to the total spectral shift.

Solvatochromic study of cycloimmonium ylids contributes to obtain information both on the electro-optical parameters of these molecules and on the contribution of each type of intermolecular interactions to the spectral shift measured in each solvent.

REFERENCES

Abe, T. 1965. Theory of the solvent effects on molecules electronic spectra. Frequency shifts. *Bull. Chem. Soc. Japan* 38:1314–1318.

Avadanei, M.I., Dorohoi, D.O., and Melniciuc-Puica, N. 2011. Ordering tendency in ternary solutions of pyridazinium ylids evidenced by electron spectroscopy. *SPIE Proc.* 8001:800112D.

Avadanei, M.I., Ivan, L.M., Nadejde, C., Creanga, D., and Dorohoi, D.O. 2015. Spectral and thermodynamical studies of iso-quinolinium carboethoxy-methylid (iQCEM) solutions with binary solvent water + ethanol. *Rev. Chim. Bucur.* 66(2):201–204.

Babusca, D. and Dorohoi, D.O. 2016. Solvent empirical scales used to study the intermolecular interactions in binary solutions of two p-aryl-pyridazinium methylids. *Spectrochim. Acta A: Mol. Biomol. Spectrosc.* 152:149–155.

Bakhshiev, N.G. 1961. Universal molecular interactions and their effect on the position of the electronic spectra of molecules in two component solutions. I. Theory (liquid solutions). *Optica I Spectroscopia* 10:379–384.

Bakhshiev, N.G. 1972. *Spectroscopia mejmoleculiarn'x vzaimodeisviax.* Nauka, Leningrad.

Bayliss, N.S. and McRae, E.G. 1954. Solvent effects in organic chemistry. Dipole forces and Franck–Condon principle. *J. Phys. Chem.* 58:1002–1006.

Benchea, A.C., Closca, V., Rusu, C.M., and Dorohoi, D.O. 2014. Electro-optical parameters in excited states of some spectrally active molecules. *Proc. SPIE* 9286:928649.

Catalan, J. 2009. Toward a generalized treatment of the solvent effect based on four empirical scales: Dipolarity (SdP, a new scale); Polarity (SP); Acidity (SA); Basicity (SB) of the medium. *J. Phys. Chem.* 113:5951–5960.

Closca, V., Ivan, L.M., and Dorohoi, D.O. 2014a. Intermolecular interactions in binary solutions of two cycloimmonium carboethoxy methylids. *Spectrochim. Acta A: Mol. Biomol. Spectrosc.* 122:670–675.

Closca, V., Zelinschi, C.B., Babusca, D., and Dorohoi, D.O. 2014b. Solvent empirical scales for electronic absorption spectra. *Uk. J. Phys.* 59(3):226–232.

Dimitriu, M., Dimitriu, D.G., and Dorohoi, D.O. 2008. Supply of the spectral shifts of each type of intermolecular interactions in binary solutions. *Optoelectron. Adv. Mater. Rapid Commun.* 2(12):867–870.

Dorohoi, D.O. 2004. *N*-Ylid spectroscopy. *J. Mol. Struct.* 704(1–3):31–43.

Dorohoi, D.O. 2006. Electric dipole moments of the spectrally active molecules estimated from the solvent influence on the electronic spectra. *J. Mol. Struct.* 92:86–92.

Dorohoi, D.O. 2007. Symmetry properties and the electronic transitions of some pyridinium ylids. *AIP Conf. Proc.* 899:364–365.

Dorohoi, D.O. 2010. About the multiple regressions applied in studying the solvatochromic effects. *Spectrochim. Acta A: Mol. Biomol. Spectrosc.* 75(3):1030e1–1030e5.

Dorohoi, D.O., Avadanei, M., and Postolache, M. 2008. Characterization of the solvation spheres of some dipolar spectrally active molecules in binary solvents. *Optoelectron. Adv. Mater. Rapid Commun.* 2(8):511–514.

Dorohoi, D.O., Dascalu, F.C., Teslaru, T., and Gheorghies, L.V. 2012. Electronic absorption spectra of two 4-aryl-pyridazinium-2,4,6-picryl benzoyl methylids. *Spectrosc. Lett.* 45:383–391.

Dorohoi, D.O., Dimitriu, D.G., Dimitriu, M., and Closca, V. 2013. Specific interactions in *N*-ylid solutions, studied by nuclear magnetic resonance and electric absorption spectroscopy. *J. Mol. Struct.* 1044:79–86.

Dorohoi, D.O., Partenie, D.H., Chiran, L.M., and Anton, C. 1994. About the electronic absorption spectra (EAS) and electronic diffusion spectra (EDS) of some pyridazinium ylids. *J. Chim. Phys. Phys. Chim. Biol.* 91:419–431.

Dorohoi, D.O. and Pop, V. 1987. The spectral shifts in visible electronic absorption spectra of some cycloimmonium ylids in ternary solution. *Analele Stiint. Universitatii "Alexandru Ioan Cuza" Iasi Ib Fiz.* XXXIII:78–86.

Dorohoi, D.O., Surpateanu, G., and Gheorghies, L.V. 1998. A spectral method to determine some molecular parameters of cycloimmonium ylids. *Balk. Phys. Lett.* 6(3): 198–203.

Dulcescu, M.M. and Dorohoi, D.O. 2009. Ternary solutions of the carbanion monosubstituted pyridazinium ylids in binary protic solvents. *Univ. Polytehnica Buchar. Sci. Bull. Ser. A Appl. Math. Phys.* 71(1):87–96.

Dulcescu, M.M., Stan, C., and Dorohoi, D.O. 2010a. Spectral investigations of the influence of the hydroxyl solvents on the intermolecular interactions in some pyridazinium ylid ternary solutions. *Rev. Chim. Bucur.* 61(12):1219–1222.

Dulcescu, M.M., Stan, C., and Dorohoi, D.O. 2010b. Spectral study of intermolecular interactions in water–ethanol solutions of some carbanion disubstituted pyridazinium ylids. *Rev. Roum. Chim.* 55(7):403 408.

Gheorghies, C., Gheorghies, L.V., and Dorohoi, D.O. 2008. Solvent influence on some complexes realized by hydrogen bond. *J. Mol. Struct.* 887:122–127.

Gheorghies, L.V., Dimitriu, M., Filip, E., and Dorohoi, D.O. 2010. Universal interactions in binary solutions studied by spectral means. *Rom. J. Phys.* 55(1–2):103–109.

Gheorghies, L.V. and Dorohoi, D.O. 2008. Intermolecular interactions in dipolar binary solutions. *Rom. J. Phys.* 53(1–2):71–77.

Homocianu, M., Airinei, A., and Dorohoi, D.O. 2011a. Intermolecular interactions in binary solutions of pyridazinium ylids studied by visible electron spectroscopy. *SPIE Proc.* 8001:80013K.

Homocianu, M., Airinei, A., Dorohoi, D.O., Olariu, I., and Fifere, N. 2011b. Solvatochromic effects in the UV Vis absorption spectra of some pyridazinium ylids. *Spectrochim. Acta A: Mol. Biomol. Spectrosc.* 82:355–359.

Johnson, A.W. 1966. *N-Ylid Chemistry*. Academic Press, New York.

Kamlet, M.J., About, J.L., and Taft, R.W. 1977. The solvatochromism comparison method 6. The pi stea scale of the solvent polarity. *J. Chem. Soc. Am.* 99:6027–6038.

Kosower, E.M. 1958. The effect of solvent on spectra I. A new empirical measure of the solvent polarity: Z-values. *J. Am. Chem. Soc.* 80:3253–3260.

Kosower, E.M. 1968. *An Introduction to Physical Organic Chemistry*. John Wiley and Sons, London.

Mataga, N. and Kubota, T. 1970. *Molecular Interactions and Electronic Spectra*. M. Decker, New York.

Mazurenko, I. 1972. Universal interactions in solutions with three component. *Opt. I Spektroskopia* XXXIII:1060–1064.

McRae, E.G. 1956. Theory of solvent effects on molecular electronic spectra. Frequency shifts. *J. Phys. Chem.* 61:562–572.

Melniciuc-Puica, N., Avadanei, M.I., Caprosu, M., and Dorohoi, D.O. 2012. Interaction energy in pairs of phthalazinium dibenzoyl methylid (PDBM)—Protic solvent molecules estimated in the limits of ternary solution model. *Spectrochim. Acta A: Mol. Biomol. Spectrosc.* 96:271–277.

Melnig, V., Humelnicu, I., and Dorohoi, D.O. 2006. Thermal dimerization kinetics of 3-(p-bromo-phenyl)-pyridazinium benzoyl methylid in solutions. *Int. J. Chem. Kinet.* 40(5):230–239.

Perov, A.N. 1980. Energy of intermediate pair interactions as a characteristic of their nature. Theory of solvate (fluoro) chromism of three components solutions. *Opt. I Spectreoscopia* 8:408–413.

Pop, V., Dorohoi, D.O., and Delibas, M. 1986. Considerations on the statistical model of intermolecular interactions in ternary solutions. *Analele Stiint. Universitatii "Alexandru Ioan Cuza" Iasi Ib Fiz.* XXXII:79–84.

Pop, V., Dorohoi, D.O., and Holban, V. 1994. Molecular interactions in binary solutions of 4-amino-phthalimide and 3-p-cumyl-pyridazinium-acetyl-benzoyl methylid. *Spectrochim. Acta A: Mol. Biomol. Spectrosc.* 50(14):2281–2289.

Reichardt, C. 2003. *Solvents and Solvent Effects in Organic Chemistry*, 3rd updated and enlarged edition. Wiley VCH, Weinheim.

Stone, A. 2013. *The Theory of Intermolecular Forces*, 2nd edition. Oxford University, Oxford.

Suppan, P. 1990. Solvatochromic shifts. The influence of the medium on the energy of electronic states. *J. Photochem. Photobiol. A: Chemistry* 50:293–330.

Surpateanu, G. and Dorohoi, D. 1977. Group electronegativity determination II. Symbiosis effect. *Analele Universitatii "Alexandru Ioan Cuza" Iasi Ib Fiz.* 23:99–102.

Surpateanu, G., Dorohoi, D., and Zugravescu, I. 1975. Group electronegativity determination I. *Analele Universitatii "Alexandru Ioan Cuza" Iasi Ib Fiz.* 21:89–90.

Wypych, G. (Ed.). 2001. *Handbook of Solvents*. Chem. Tech. Publishing, Toronto.

Zugravescu, I. and Petrovanu, M. 1976. *N-Ylid Chemistry*. McGraw Hill, International Book Company, New York.

4 Relevance of the Molecular Descriptors for the Modeling/ Discrimination of Amphetamines Using Artificial Neural Network

Steluta Gosav

CONTENTS

4.1 INTRODUCTION

Amphetamines, generally referred to as phenylethylamines (Figure 4.1), are the class of compounds with the largest number of individual substances on the illicit drug market. They are abused for their stimulant and/or hallucinogenic effect. Amphetamine-type stimulants (ATS) are part of the psychostimulant group of drugs and include meth/amphetamine, ecstasy, cocaine, and some pharmaceuticals, that is, dexamphetamine and ritalin. All stimulants work by increasing dopamine levels in the brain—dopamine is a brain chemical (or neurotransmitter) associated with pleasure, movement, and attention. The therapeutic effect of stimulants is achieved by slow and steady increases of dopamine, which are similar to the natural production

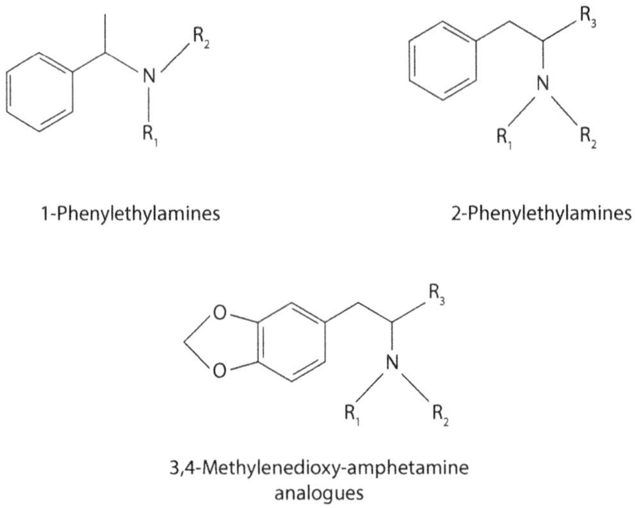

1-Phenylethylamines

2-Phenylethylamines

3,4-Methylenedioxy-amphetamine
analogues

FIGURE 4.1 Illicit amphetamines.

of the chemical by the brain. ATS stimulate central nervous system activity, producing euphoria, a sense of well-being, wakefulness, and alertness. However, ATS use is associated with a range of potentially negative health consequences such as an increased heart rate, blood pressure and body temperature, sleeplessness, and reduced appetite. The increases in blood pressure and heart rate can affect organs and can contribute to stroke, heart problems, and kidney failure (Miczek and Tidey 1989, Karch 1998).

MDMA or 3,4-methylenedioxymethamphetamine and other compounds of similar chemical structure and action are called hallucinogenic amphetamines (Figure 4.1). Their action is the result of a combination of the stimulating effect of ATS amphetamine and the hallucinogenic effect of mescaline.

The action of MDMA can best be defined by the term "entactogenic," describing a state of a slight distortion of reality, breaking down of emotional barriers, sense of joy, intimacy, closeness to other people, good mood, acceptance of the self and of the world, and a heightening of one's perception of the surroundings (Robson 1997, Nasiadka et al. 2002). The evocation of a euphoric state, the entactogenic effect, and the easy availability of the drug mean that they are becoming more and more popular, especially among young people. In an attempt to circumvent the laws controlling drugs of abuse, new chemical structures are very frequently introduced on the black market. They are obtained by slightly modifying the controlled molecular structures by adding or changing substituents at various positions on the banned molecules. As a result, no substance similar to those forming a prohibited class may be used nowadays, even if it has not been specifically listed. Therefore, reliable, fast, and accessible expert systems capable of modeling and then identifying similarities at the molecular level are highly needed for epidemiological, clinical, and forensic purposes.

In the last few years, the application of the artificial neural networks (ANNs) in different fields of spectroscopy, for example, infrared (IR) (Gosav et al. 2005, 2006) or mass spectrometry (MS) (Gosav et al. 2008, 2011), has become an efficient tool for structure identification/classification. An ANN is a biologically inspired computer program designed to learn from data in a manner of emulating the learning pattern in the brain. Most ANN systems are very complex high-dimension processing systems. ANNs are good recognizers of patterns and robust classifiers, with the ability to learn and generalize from examples in order to produce meaningful solutions to problems even when the input data contain errors or are incomplete. A major advantage of ANNs compared to multivariate statistical methods is that they do not require rigidly structured experimental designs and can map functions using incomplete data (Massart et al. 1997). ANNs are widely used in pharmacology, especially in SARs/QSARs (structure activity relationships/quantitative structure activity relationships) (Deeb and Hemmateenejad 2007, Prakash et al. 2013) and STR/QSTR (structure toxicity relationships/quantitative structure toxicity relationships) studies (Gao et al. 2003, Gosav et al. 2007, Cheng and Vijaykumar 2012). QSAR/QSTR correlates physicochemical parameters, that is, topological parameters, molecular properties (logP, hydrogen acceptor, hydrogen donor, molar refractivity, etc.), geometrical descriptors (GD), weighted holistic invariant molecular (WHIM) descriptors, etc., of compounds with biological/toxicological activity. ANNs have been shown to be an effective tool to establish this type of relationship and predict the activities of new compounds based on their molecular descriptors.

A molecular descriptor is the final result of a logical and mathematical procedure, which converts chemical information from a symbolic representation, that is, three-dimensional Euclidean representation, two-dimensional representations based on the graph theory, etc., of the molecule into a useful numeric value which represents the theoretical descriptor (Todeschini and Consonni 2000). According to the kind of used molecular representation and to the defined algorithm for calculation of the theoretical descriptor, there are simple molecular descriptors (0D and 1D-descriptors), that is, constitutional descriptors (CDs), functional group counts (FGC) and atom-centered fragments (ACF), topological descriptors (TD) (2D-descriptors) derived from algorithms applied to a molecular graph, and GD (3D-descriptors) obtained directly from the (x, y, z) atomic coordinates of the molecule atoms. As the molecular descriptors contain important information regarding the molecular structure of a chemical compound, they are very useful in describing and classifying the structure, in relating the structure to properties, and predicting chemical and biological properties.

The goal of this chapter is to develop an expert system which is able to identify the potential biological activity of new substances having a molecular structure similar to illicit amphetamines. To this purpose, we have designed 14 ANNs, which have been trained to classify amphetamines according to their toxicological activity (stimulant amphetamines or hallucinogenic amphetamines) and to distinguish them from nonamphetamines based on their molecular descriptors, that is, CDs, FGC, ACF, TD, walk and path counts (WPC), connectivity indices (CI), information indices (II), Burden eigenvalues (BE), edge adjacency indices (EAI), GD, RDF (radial distribution function), WHIM descriptors, GETAWAY (GEometry,

Topology, and Atom-Weights AssemblY), and 3D-MoRSE (3D-molecule representation of structures based on electron diffraction) descriptors. Such a system is essential for testing new molecular structures for epidemiological, clinical, and forensic purposes.

The efficiency with which each network identifies the class identity of an unknown sample was evaluated by calculating several figures of merit. The results of the comparative analysis are presented.

4.2   BACKGROUND OF ANN MODEL

The ANN is a computer-based system derived from a simplified concept of the brain in which a number of nodes, called processing elements (neurons), are interconnected in a network-like structure. The nodes operate in parallel and can be trained to map a set of input patterns into a set of output patterns. Each processing element has the basic functionality of a biological neuron: to receive signals, to sum the signals, to transform the signals with an activation function, and to produce a signal (Figure 4.2) that is passed into other elements (Zupan and Gasteiger 1999).

The most common architecture of an ANN is the multilayer feed-forward network, in which the nodes are organized in three types of layers: a layer of input nodes, one or more layers of hidden nodes, and a layer of output nodes (Figure 4.3). Each node in one layer is connected to all the nodes of the next one. Neurons in the input layer only act to distribute the input signals X_i ($i = 1, 2, \ldots, n$) to neurons in the hidden layer. Each neuron j (Figure 4.2) in the hidden layer sums up its input signals x_i after weighting them with the strengths of the respective connections w_{ji} from the input layer obtaining thus the whole input signal, Net_j.

$$Net_j = \sum_{i=1}^{m} w_{ji} x_i \qquad (4.1)$$

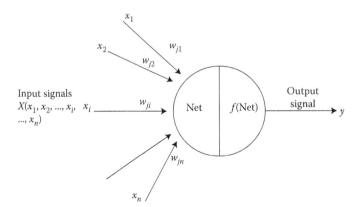

FIGURE 4.2   The neuron j with its input and output signals.

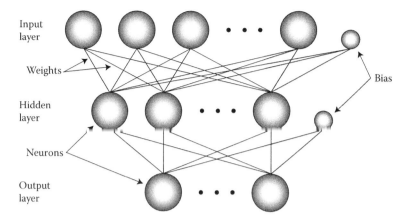

FIGURE 4.3 The architecture of an ANN with three layers.

Then, the neuron j computes its output y_j as a function f, that is, the activation function, of this sum.

$$y_j = f\left(\sum_{i=1}^{m} w_{ji} x_i\right) \tag{4.2}$$

The output of neurons in the output layer is computed similarly.

The advantage of ANN is that it is possible to "train" the network. It learns by modifying the values of the weights in a well-defined manner, described by the learning rules. The general type of learning is supervised learning, which requires knowledge of the desired responses, that is, target (t), to input signals and incorporates an algorithm to apply the learning rule.

The aim is to minimize the errors between the target and computed output values by adjusting the weights of the connections beginning from the output layer to the input layer, iteratively. This type of training is applied by the backpropagation algorithm which uses the gradient descent method, a method which decides how to change the weight in order to get closer to the global minimum of the error surface (Zupan and Gasteiger 1999).

The weight change (correction) $\Delta w_{ji}^{\ell(new)}$ of the connection between the neuron i in the $(\ell - 1)$ layer and the neuron j in the ℓ layer is given by the expression (Gosav 2008)

$$\Delta w_{ji}^{\ell(new)} = \eta \delta_j^{\ell} y_i^{\ell-1} + \mu \Delta w_{ji}^{\ell(previous)} \tag{4.3}$$

where $\Delta w_{ji}^{\ell(previous)}$ is the weight change performed in the previous training cycle, η is the learning rate, $y_i^{\ell-1}$ is the output of neuron i in the layer $\ell - 1$, μ is the "momentum" coefficient, and δ_j^{ℓ} is the error generated by the neuron j in the layer ℓ.

The term δ_j^ℓ depends on whether neuron j is an input neuron or a hidden neuron. For output neurons

$$\delta_j^{out} = \left(t_j - y_j^{out}\right) y_j^{out} \left(1 - y_j^{out}\right) \tag{4.4}$$

and for hidden neurons

$$\delta_j^\ell = \left(\sum_{k=1}^{r} \delta_k^{\ell+1} w_{kj}^{\ell+1}\right) y_j^\ell \left(1 - y_j^\ell\right) \tag{4.5}$$

where y_j^{out} and y_j^ℓ are the outputs of neuron j in the output layer and layer ℓ, respectively. t_j is the target output for neuron j and r is the number of neurons in the layer $(\ell + 1)$.

By adding the "momentum" coefficient to Equation 4.3, an enhancement of the speed of training was produced by the fact that the previous weight change $\Delta w_{ji}^{\ell(previous)}$ effectively influences the new weight change $\Delta w_{ji}^{\ell(new)}$ (Al Shamisi et al. 2011).

4.3 DESIGNING AND PROGRAMMING OF ANN MODEL

Generally, the procedure for the designing of ANN models follows two basic steps: building the database, that is, collecting and preprocessing data, and developing the ANN model, that is, building the network, training, and testing the ANN model.

4.3.1 BUILDING THE DATABASE

The database consists of 145 compounds representing forensic substances such as drugs of abuse (mainly central stimulants, hallucinogens, sympathomimetic amines, narcotics, and other potent analgesics), precursors, or derivatized counterparts. These molecules were represented in 3D coordinates, using *HyperChem* (version 6.03) molecular modeling software. Their geometries were fully optimized using the semiempirical quantum chemical method AM1. The files containing the optimized molecules are used by Dragon 5.5 software as inputs in order to compute a number of 1419 molecular descriptors representing 14 categories of descriptors.

After data collection, two data preprocessing procedures were performed in order to train the ANNs more efficiently. These procedures are the following: solve the problem of constant (or almost constant) data and normalize the data. Thus, from a total number of 1419 molecular descriptors, after the elimination of the descriptors with constant or almost constant values for all compounds, 1153 descriptors remained. The dimension of the data matrix is of $n \times m$, where $n = 145$ and $m = 1153$ are the numbers of compounds and descriptors, respectively. Also, the data were preprocessed using mean centering so that they could be compared to each other on the same scale and, while all data have the same importance in the analysis. All categories of descriptors and the number of descriptors corresponding to each category are shown in Table 4.1. A detailed description of the molecular descriptors, their mathematical representation, and chemical meaning is presented in Todeschini and Consonni (2000).

TABLE 4.1

Computed Theoretical Categories of Descriptors and the Number of Descriptors Corresponding to Each Category

Type of Descriptors	Code	Number of Descriptors
0D descriptor		
Constitutional descriptors	CD	40
1D descriptors		
Functional group counts	FGC	31
Atom-centered fragments	ACF	44
2D descriptors		
Topological descriptors	TD	90
Walk and path counts	WPC	46
Connectivity indices	CI	33
Information indices	II	47
Edge adjacency indices	EAI	106
Burden eigenvalues	BE	64
3D descriptors		
Geometrical descriptors	GD	48
RDF descriptors	RDF	148
3D-MoRSE descriptors	MoRSE	160
WHIM descriptors	WHIM	99
GETAWAY descriptors	GETAWAY	197
Total		1153

4.3.2 Development of ANN Models

In order to develop an ANN, it is necessary to specify the constitution of the input and output vectors, the architecture and the algorithm, the training stage, and the performances of the network. In this chapter, the ANN systems were designed with four layers, that is, an input layer, two hidden layers, and an output layer.

All networks have three outputs, distinguishing between stimulant amphetamines (class code M), hallucinogenic amphetamines (class code T), and nonamphetamines (class code N). The difference among them is that they were built with a different number of input variables representing each category of molecular descriptors.

Then, the total number of ANN models is 14: a *0D-ANN* model which has the CDs as inputs, two *1D-ANN* networks having as input variables the FGC and the ACF, six *2D-ANN* models which use as inputs the descriptors 2D, and five *3D-ANN* models with input variables representing each category of the descriptors 3D (Table 4.1). All ANN models were trained using the backpropagation algorithm with momentum and adaptive learning rate, and the sigmoidal function was adopted as the transfer function (activation).

The ANN model includes two stages, that is, the training stage and the validation stage. In the training stage, the neural network learns the potential relationship between the input data, that is, molecular descriptors, and the class assignments (M, T, and N classes) from a training set. The learning process is continued until the network converges, that is, the root mean square error (RMSE) of training drops below a target error. RMSE measures the variation of predicated values (y_j) around the target (t_j), its expression being the following:

$$\text{RMSE} = \sqrt{\frac{1}{n}\sum_{j=1}^{n}(t_j - y_j)^2} \tag{4.6}$$

where n is the number of compounds from the training set. Once the training process is complete, a relationship between the input and output data is established and the network is ready for validation.

Classification experiments were carried out on 145 molecules, out of which 14 are stimulant amphetamines, 7 are hallucinogenic amphetamines, and 124 are non-amphetamines (counterparts). The training set for the ANNs consists of 29 samples: 6 stimulant amphetamines, 5 hallucinogenic amphetamines, and 18 nonamphetamines. The remaining 116 samples were included in the validation set. The stimulants of the training set are benzphetamine, methamphetamine, β-phenylethylamine, α-phenylethylamine, fenproporex, and N-methyl-α-phenylethylamine. The hallucinogens are 3,4-methylenedioxyamphetamine, 3,4-methylenedioxymethamphetamine, 3,4-methylenedioxy-N-ethylamphetamine, N-methyl-1-(3,4-methylenedioxyphenyl)-2-butanamine, and 1-(3,4-methylenedioxyphenyl)-2-butanamine. The nonamphetamines of the training set were selected randomly from the database: bemegride, β-butyrolactone, cadaverine, and its HFB-derivate, codeine and its pentafluoropropionic (PFP)-derivate, caffeine, γ-butyrolactone, the trimethylsilyl (TMS)-derivate of γ-hydroxy butyric acid, the TMS-derivate of γ-hydroxy valeric acid, γ-valerolactone, 4-iodo-2,5-dimethoxyamphetamine, ketamine, nicotine, nicotamide, piracetam, prolintane, and putrescine.

All ANN systems were validated using all the samples from the database. The method was full cross-validation, as the number of samples in the database is relatively small. The results of validation were quantified by computing of several figures of merit such as the rate of true positives (TP), of true negatives (TN), of false positives (FP), of false negatives (FN), of classification (C), and of correctly classified samples (CC).

4.3.3 Programming of ANN Model

An ANN was developed using an "*in-house*" software which was written using Neural Network Toolbox, MATLAB® 7.0.0 (MATLAB 7.0.0 2004). The Neural Network Toolbox contains the MATLAB tools for designing, implementing, visualizing, and simulating neural networks. Figure 4.4 shows us the procedural steps to develop the ANN. The ANN program starts by loading preprocessed data from Excel files, that is, training and validation data. The next step in building the ANN

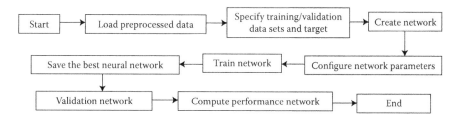

FIGURE 4.4 Flow chart for developing ANN model using MATLAB.

model was the generation of matrices of training and validation data sets and the establishing of the target vector. Then, the architecture of the neural network was defined, that is, the number of layers and the dimension of each layer. Also, the initial weights of the connections and the biases were randomized and the activation function of neurons was established.

The feed-forward backpropagation network object, *net,* was created using the built-in *nnt2ff* MATLAB function:

$$net = nnt2ff(minmax(p), \{w1w2w3\}, \{b1b2b3\}, \{f1f2f3\})$$

where the argument *p* is the training set, *w*1, *w*2, and *w*3 are the matrices of initial weights of the connections linking the neurons of neighboring levels. The arguments *b*1, *b*2, and *b*3 representing the biases for layers, and *f*1, *f*2, and *f*3 are the transfer functions for neurons of the active layers, that is, the hidden layers and the output one. The neural network uses implicitly as training function the "*traingdx*" function that updates weight and bias values according to gradient descent momentum and an adaptive learning rate.

The neural network was next configured as follows:

$$net.trainParam.epochs = 700;$$
$$net.trainParam.goal = 0.0001;$$

where "*trainParam.epochs*" denotes the maximum number of training epochs and "*trainParam.goal*" represents the mean-squared error goal (target error), set here to be 0.0001. All ANN networks have been programmed to stop the training process when the RMSE of training drops below the target error (Figure 4.5).

In order to perform the training of the network, the "*train()*" *MATLAB* function was applied:

$$[net, tr, y] = train(net, p, t);$$

where the argument *net* is the network object defined above, *p* is the training data, and *t* is the target vector. The results of application of the *train()* function are the following: the new network object *net* with the matrices of adjusted weights, *tr* is the number of training cycles (epochs) and *y* is the predicted output vector.

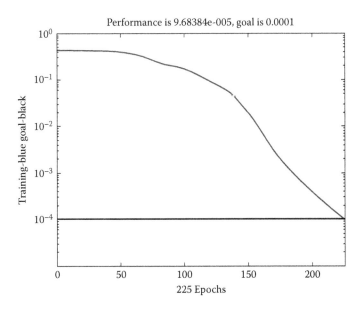

FIGURE 4.5   RMSE error of training versus the number of epochs for *0D-ANN* network.

The best ANN model for each type of descriptors (inputs) was obtained by repeating two previous steps several times, that is, definition network structure and train network. Then, the best ANN model was saved and used for the validation process. At the validation stage, unseen data are exposed to the model. The testing simulation process is called with the statement

$$[Y] = sim(net, z1);$$

where the argument *net* is the best neural network and *z1* denotes the validation data, that is, matrices 145 × number of input variables. The simulation process generates the predicted output data *[Y]*. Before the validation process, it performed a randomized permutation of validation data.

4.4   RESULTS AND DISCUSSION

In order to offer a complete image of the validation results, the figures of merit for all ANN models, that is, *0D-ANN, 1D-ANN, 2D-ANN,* and *3D-ANN* networks, specialized in the automated classification of illicit amphetamines are presented in Tables 4.2, 4.4, and 4.5. Taking into account that the ANN models have to recognize psychotropic substances, the most important figure of merit to be analyzed is the TP rate. The value of this parameter gives us real information regarding the sensitivity of the network only if that ANN model classifies (correctly or incorrectly) all positives. It is important to mention that the ANN models built in this chapter classify all amphetamines (stimulants and hallucinogens) from the database excepting a *3D-ANN* network, that is, the GETAWAY_ANN model, which fails to classify a hallucinogenic amphetamine.

TABLE 4.2

Architecture and the Validation Parameters for *0D-ANN* and *1D-ANN* Networks

	CD_ANN	FGC_ANN	ACF_ANN
Network Structure	40-24-12-3	31-20-12-3	44-22-12-3
TP (%)	100	100	100
TN (%)	89	84	81
FP (%)	11	16	19
FN (%)	0	0	0
C (%)	97	90	95
CC (%)	91	87	84

4.4.1 Performance Analysis of 0D-ANN and 1D-ANN Models

A first and very encouraging observation is that *0D-ANN* and *1D-ANN* networks (see Table 4.2) have a very good sensitivity, that is, the TP rate is equal to 100%, meaning that all positives (amphetamines) are recognized as such. Also, it can be seen that the selectivity of these networks is good (TN = 89% for CD–ANN model, TN = 84% for FGC–ANN, and TN = 81% for ACF–ANN model, respectively), the TN rate for *0D-ANN* network being greater than that for the *1D-ANN* networks. This result is very important, as the modeling of the class of nonamphetamines is difficult, the compounds belonging to the N class having very different molecular structures.

A second remark is related to the fact that the rates of C and of CC for the *0D-ANN* and *1D-ANN* networks have good values (C = 97% and CC = 91% in the case of CD–ANN model, C = 90% and CC = 87% in the case of FGC–ANN model, and C = 95% and CC = 84% in the case of ACF–ANN model, respectively), the best performing network being the CD–ANN model. Therefore, one may conclude that the CDs (Table 4.3) bring more important information to the ANN model than 1D descriptors, that is, fragment group counts and ACF.

By comparing the validation results of the 1D-ANN networks (Table 4.2), one can see that the ACF_ANN model is slightly more efficient than FGC_ANN model. A lesser apparent selectivity of the ACF_ANN network (TN = 81% in the case of ACF_ANN network and TN = 84% in the case of FGC_ANN model) is compensated by a higher classification rate C (C = 95% in the case of ACF–ANN model and C = 90% in the case of FGC–ANN model).

4.4.2 Performance Analysis of 2D-ANN Models

Analyzing the figures of merit for the *2D-ANN* models (Table 4.4), one could observe that the identifying of positives (especially of hallucinogens) constitutes a challenge in the case of the WPC_ANN, II_ANN, and BE_ANN networks, their TP rate being very small, that is, 67%. In the case of the other three models,

TABLE 4.3
List of Constitutional Descriptors

Description	Symbols
Molecular weight, average molecular weight	MW, AMW
Sum of atomic van der Walls volumes, Sanderson electronegativities, and polarizabilities (scaled on C atom)	Sv, Se, and Sp
Sum of Kier–Hall electrotopological states	Ss
Mean atomic van der Waals volume, Sanderson electronegativities, and polarizabilities (scaled on C atom)	Mv, Me, and Mp
Mean electrotopological state	Ms
Number of atoms, of non-H atoms	nAT, nSK
Number of bonds, of non-H bonds and of multiple bonds	nBT, nBO, and nBM
Sum of conventional bond orders (H-depleted)	SCBO
Aromatic ratio	ARR
Number of rings and circuits	nCIC, nCIR
Number of rotatable bonds	RBN
Rotatable bond fraction	RBF
Number of double and aromatic bonds	nDB, nAB
Number of H, C, N, O, F, Cl, and I atoms	nH, nC, nN, nO, nF, nCl, and nI
Number of heavy and halogen atoms	nHM, nX
Number of 5-, 6-, 8-, 9-, 10-, 11-, and 12-membered rings	nR05, nR06, nR08, nR09, nR10, nR11, and nR12
Number of benzene-like rings	nBnz

TABLE 4.4
Architecture and the Validation Parameters for *2D-ANN* Networks

Network	TD_ANN	WPC_ANN	CI_ANN	II_ANN	EAI_ANN	BE_ANN
Structure	90-40-12-3	46-22-12-3	33-16-12-3	47-22-12-3	106-50-12-3	64-30-12-3
TP (%)	100	67	95	67	95	67
TN (%)	81	68	69	85	68	82
FP (%)	19	32	31	15	32	18
FN (%)	0	33	5	33	5	33
C (%)	95	98	95	97	93	96
CC (%)	84	68	73	82	72	80

the sensitivity is very high, the TP rate having the following values: 95% for the EAI_ANN and CI_ANN networks and 100% for the TD_ANN network, respectively. It is important to mention that in the case of EAI, the ANN model classifies a stimulant as hallucinogen and, in the case of CI, the ANN model identifies six hallucinogens as stimulants.

Consequently, these two types of descriptors fail to classify the amphetamines according to their toxicological activity, that is, stimulant amphetamines and hallucinogenic amphetamines, having a great modeling power for amphetamine class (M and T). In conclusion, the most sensitive *2D-ANN* network is the TD_ANN model with TD as input variables. For the same network, the rate of CC has the highest value (84%) in comparison with other *2D-ANN* networks.

Regarding the selectivity of the 2D_ANN models, it is worth noticing that even if the II_ANN and BE_ANN models have the value of TN rate, that is, 83% for II_ANN model and 82% for BE_ANN model, slightly higher than that for TD–ANN model, that is, TN = 81%, this advantage of the former models is canceled due to their very low sensitivity (TP = 67%). Following the same reasoning, the value of the classification rate (C) slightly lesser for TD–ANN network (C = 95%) than for the WPC_ANN, II_ANN, and BE_ANN networks (C = 98%, C = 97%, and C = 96%, respectively) does not diminish the efficiency of the former network due to its excellent sensitivity (TP = 100%). Finally, the TDs show a stronger modeling/discrimination power than other 2D descriptors.

4.4.3 Performance Analysis of 3D-ANN Models

The validation results presented in Table 4.5 reveal to us that the true positive rate has the maximum value, that is, TP = 100%, in the case of WHIM_ANN network and the rate of the correctly classified sample has the greatest value, that is, CC = 85%, for the same network. Regarding the TN and C rates, one can see that their values are good in the case of WHIM_ANN network, that is, TN = 82% and C = 91%. Based on the overall characterization, it ensures that the WHIM_ANN network performs better than the other 3D networks. Therefore, the WHIM descriptors are the 3D descriptors that bring the most relevant information for the class identity assignment. Besides the 2D-ANN networks, there are also several 3D-ANN networks, that is, WHIM_ANN, RDF_ANN, and GETAWAY_ANN models, which cannot make a very good recognition of amphetamines according to their toxicological activity. Thus, in the case of RDF descriptors, the ANN model classifies a hallucinogen as

TABLE 4.5
Architecture and the Validation Parameters for *3D-ANN* Networks

Network Structure	WHIM_ANN 99-46-12-3	RDF_ANN 148-70-12-3	MoRSE_ANN 160-80-12-3	GETAWAY_ANN 197-100-12-3	GD_ANN 48-20-12-3
TP (%)	100	71	76	90	86
TN (%)	82	84	79	80	69
FP (%)	18	16	21	20	31
FN (%)	0	29	24	10	14
C (%)	91	94	92	92	93
CC (%)	85	82	79	81	72

the stimulant and, in the case of GETAWAY descriptors, the ANN model identifies two hallucinogens as stimulants.

A special situation is the case of WHIM descriptors because the ANN model identifies a hallucinogen as a stimulant and, inversely, a stimulant as a hallucinogen. For this reason, one may draw the conclusion that the WHIM descriptors are more appropriate for modeling the whole class of amphetamines, that is, stimulants and hallucinogens.

In order to compare the performances of the most efficient ANN networks from each category, that is, *0D-ANN, 1D-ANN, 2D-ANN,* and *3D-ANN* networks, we have plotted in Figure 4.6 the TP and the TN rates and in Figure 4.7 the C and the CC rates for the following networks: CD–ANN, ACF_ANN, TD_ANN, and WHIM_ANN. Figure 4.6 confirms the fact that these ANN networks are very fit-for-the-purpose. Taking into account the legal consequences, in forensic practice, no positive sample should "escape" unidentified during the screening process. Indeed, the above-mentioned networks have a very good sensitivity (TP rate of 100%). Regarding the selectivity, there is a challenge for the classification of negatives (nonamphetamines).

The networks are not as efficient as in the case of the positives, showing that the descriptors, regardless of their category, do not allow the network to have a sufficient modeling power for the class of nonamphetamines. Thus, one can see (Figure 4.6) that the ACF_ANN, TD_ANN, and WHIM_ANN networks have nearly the same TN value (TN $\cong$ 80%), the ANN network with the highest value being the structural CD–ANN network (TN = 89%). Moreover, the CD–ANN network has the best values for the C (97%) and CC (91%) rates (Figure 4.7) as well. In conclusion, one can state that the structural CD–ANN network is the most efficient expert system for indicating the class identity of illicit amphetamines.

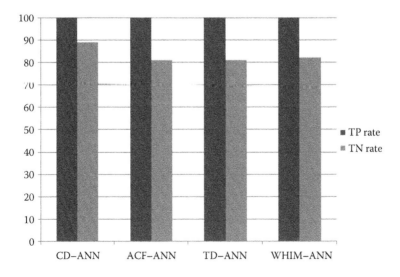

FIGURE 4.6 Comparison between the TP and the TN rates for CD–ANN, ACF_ANN, TD_ANN, and WHIM_ANN networks.

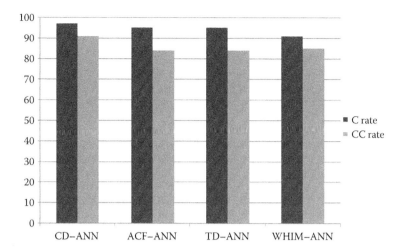

FIGURE 4.7 Comparison between the C and the CC rates for CD–ANN, ACF_ANN, TD_ANN, and WHIM_ANN networks.

4.5 CONCLUSIONS

Amphetamines do have analogues with legal use. When a new compound is to be synthesized, it is very useful to know beforehand if it will have a biological activity similar to the legal amphetamines or if its toxicity is similar to the illicit amphetamines and thus the compound will not have extensive use. The costs related to the synthesis, clinical, and toxicological tests can be avoided if the molecular structure is first analyzed with the structural system presented in this chapter. If the new molecular structure is classified by the structural CD–ANN network as positive, then, it is known that the compound most probably has the same biological activity as the illegal stimulant or hallucinogenic amphetamines, and thus, that structure can be eliminated from the research and/or production objectives. If it is classified as "nonamphetamine," then, one can conclude that although it has a molecular structure similar to the illegal amphetamines, it is not likely to have the same toxicity as this class of forbidden compounds. The chances to be useful for legal pharmaceutical purposes are then significant and further research and testing effort are justified.

REFERENCES

Al Shamisi, H.M., Assi, A.H., and Hejase, H.A.N. 2011. Using MATLAB to develop artificial neural network models for predicting global solar radiation in Al Ain City—UAE. In *Engineering Education and Research Using MATLAB*, ed. Ali, H., Assi, 219–238. InTech, Rijeka.

Cheng, F. and Vijaykumar, S. 2012. Applications of artificial neural network modeling in drug discovery. *Clin. Exp. Pharmacol. Physiol.* 2(3). http://www.omicsonline.org/applications-of-artificial-neural-network-modeling-in-drug-discovery-2161-1459.1000e113.pdf

Deeb, O. and Hemmateenejad, B. 2007. ANN–QSAR model of drug-binding to human serum albumin. *Chem. Biol. Drug Des.* 70(1):19–29.

Gao, D.W., Wang, P., and Liang, H. 2003. A study on prediction of the bio-toxicity of sub-stituted benzene based on artificial neural network. *J. Environ. Sci. Health, Part B* 38:571–579.

Gosav, S. 2008. *Studies Concerning the Identification of Illicit Amphetamines Using Spectral and Artificial Intelligence Methods (Romanian)*. Tehnopress, Iasi, Romania.

Gosav, S., Dinica, R., and Praisler, M. 2008. Choosing between GC–FTIR and GC–MS spectra for an efficient intelligent identification of illicit amphetamines. *J. Mol. Struct.* 887(1–3):269–278.

Gosav, S., Praisler, M., and Birsa, M.L. 2011. Principal component analysis coupled with arti-ficial neural networks—A combined technique classifying small molecular structures using a concatenated spectral database. *Int. J. Mol. Sci.* 12:6668–6684.

Gosav, S., Praisler, M., and Dorohoi, D.O. 2007. ANN expert system screening for illicit amphetamines using molecular descriptors. *J. Mol. Struct.* 834–836:188–194.

Gosav, S., Praisler, M., Dorohoi, D.O., and Popa, G. 2005. Automated identification of novel amphetamines using a pure neural network and neural networks coupled with principal component analysis. *J. Mol. Struct.* 744–747:821–825.

Gosav, S., Praisler, M., Van Bocxlaer, J., De Leenheer, A.P., and Massart, D.L. 2006. Class identity assignment for amphetamines using neural networks and GC–FTIR data. *Spectrochim. Acta Part A* 64:1110–1117.

Karch, S.B. 1998. *Drug Abuse Handbook*. CRC Press/Taylor & Francis, New York.

Massart, D.L., Vandeginste, B.G., Buydens, L.M.C., De Jong, S., Lewi, P.J., and Smeyers-Verbeke, J. 1997. *Handbook of Chemometrics and Qualimetrics: Part B*. Elsevier, Amsterdam, the Netherlands.

MATLAB 7.0.0 software. 2004. The Math Works Inc., Natick, MA, USA.

Miczek, K.A. and Tidey, J.W. 1989. Amphetamines: Aggressive and social behavior. In *Pharmacology and Toxicology of Amphetamine and Related Designer Drugs*, eds. K. Asghar and E. De Souza, 68–101. National Institute on Drug Abuse, Washington.

Nasiadka, K., Rutkowska, A., and Brandys, J. 2002. Hallucinogenic amphetamines. *Probl. Forensic Sci.* 52(LII):64–86.

Prakash, O., Khan, F., Sangwan, R.S., and Misra, L. 2013. ANN–QSAR model for virtual screening of androstenedione C-skeleton containing phytomolecules and analogues for cytotoxic activity against human breast cancer cell line MCF-7. *Comb. Chem. High Throughput Screen.* 16(1):57–72.

Robson, P. 1997. *Drugs*. Medical Practical Publishing, Krakow.

Todeschini, R. and Consonni, V. 2000. *Handbook of Molecular Descriptors*. Wiley-VCH, Weinheim.

Zupan, J. and Gasteiger, J. 1999. *Neural Networks in Chemistry and Drug Designing*. Wiley-VCH, Weinheim.

5 Methods for Evaluation of Light Double Refraction in Transparent Uniax Anisotropic Media

Ecaterina-Aurica Angheluţă
and Mihai-Daniel Angheluţă

CONTENTS

Abstract

The chapter brings forward multiple ways to determine the birefringence of a uni-axial anisotropic medium from the study of channeled spectra. Analyzed anisotropic layers, of various thicknesses, useful for introducing an optical path variation, comparable to the wavelength of the used radiation were analyzed in this chapter. The methods which neglect the birefringence dispersion to those which take this aspect into account were compared. Finally, the importance of knowing the birefringence in the development of several applications, such as the compensatory slides, is underlined.

5.1 INTRODUCTION

Solid substances can be classified by their microscopic structure into two groups, amorphous substances and crystalline substances, which, in turn, can be categorized as having high-, medium-, or low-symmetry crystalline structures.

High-symmetry crystalline substances (Dumitraşcu et al. 2006) have at least two $n \geq 3$ order axes. As the electric polarization vector $\vec{P}$ and the electric field intensity vector $\vec{E}$ are coaxial and $\varepsilon_0 \vec{E} + \vec{P} = \vec{D}$, the electric field intensity vector $\vec{E}$, electric induction vector $\vec{D}$, and electric polarization vector $\vec{P}$ are colinear.

Medium-symmetry crystalline substances are those which have a single $n \geq 3$ order symmetry axis. In this case, the $\vec{P}$, $\vec{E}$, and $\vec{D}$ vectors are not colinear anymore because the angle between the polarization and the electric field intensity vectors depends on the orientation of the latter relative to the main axes of the crystalline network.

In most cases, in anisotropic media, the propagation direction of the constant phase surfaces differs from the optical radiation radius (the direction of the energy transport) (Pop Valer 1988).

Low-symmetry crystalline substances have no $n \geq 3$ order symmetry axis.

The use of polarized light is a remarkable instrument in the optical study of anisotropic media. For transversal waves, there are two directions, x and y, on which one can have different amplitudes for the wave function and different relative phases.

A given relation between the amplitude and the phases of the two independent transversal fields is termed as the state of polarization. Different states of polarization may not interact with the substance in the same manner. Asymmetrical interactions take place and as a result the polarization state is modified. This event has key consequences. Studying the effect of an incident beam, with an unknown state of polarization, on certain well-known materials, we are able to determine the polarization state. Reciprocally, by measuring the modification a given material brings to a known polarization state, we can obtain information regarding its properties.

5.2 UNIAXIAL ANISOTROPIC MEDIA

In the case of isotropic substances (Giancoli 2004), the propagation speed of a linearly polarized wave is independent of the orientation of the vibration plane relative to the propagation direction. As a consequence, isotropic media do not modify the polarization state of the waves which pass through them.

Uniaxial anisotropic substances feature a privileged direction, termed optical axis, along which the propagation speed of the electromagnetic wave does not depend on the orientation of the electric field intensity vector; the light propagates in a similar manner as in an isotropic medium.

The optic axis is generally parallel to the highest-ranked symmetry axis.

For any other propagation direction, the uniaxial medium features two different refractive indices: n_o—the ordinary refractive index and n_e—the extraordinary refractive index of the medium. On a direction different from the optic axis, there are two linearly polarized waves which preserve their polarization state and propagate with different speeds, corresponding to the two different refractive indices.

The distinction between the ordinary and the extraordinary radii is that the former obeys the rule of the refraction phenomenon, while the latter does not (Halliday and Resnick 1975).

As a result, when light passes from an isotropic medium through a uniaxial anisotropic medium, a double-refraction phenomenon, corresponding to the two radii described above, takes place. It can be demonstrated that in the case of uniaxial anisotropic media, the vibration plane of the ordinary wave is perpendicular to the direction of the optic axis, while the vibration plane of the extraordinary wave contains the optic axis. The birefringence of uniaxial media is defined as the difference between the ordinary refractive index and the extraordinary refractive index ($\Delta n = n_e - n_o$).

The optic sign is given by the birefringence sign, specifically, if $\Delta n > 0$, *the crystal is uniaxial positive*; if $\Delta n < 0$, *the crystal is uniaxial negative*.

Quartz is a uniaxial positive substance (Cone 1990).

5.3 CHANNELED SPECTRA

Channeled spectra are the result of the interference of dephased waves in anisotropic substances (Chirica et al. 1985, Pop et al. 1987). In order to explain the theoretical principle for obtaining a channeled spectrum (Delibas and Dorohoi 1999), one will consider a system consisting of two polarizing filters: polarizer P (P//Ox) and analyzer A (A//Oy), between which there is an anisotropic slide (uniaxial crystal), cut parallely to the optic axis and orientated perpendicularly to the propagation direction of an equienergetic (the energy transported by the beam is equally distributed onto harmonic components) white light beam of parallel rays. Let us suppose that the propagation direction of the beam of parallel rays Oz is coincident with the Ob axis, of the main axis system attached to the anisotropic medium (Figure 5.1).

In the anisotropic slide, the electric field intensity vector is decomposed into two, reciprocally perpendicular components, with their directions of action parallel to the main Oa and Oc axes.

We denote by α the azimuth of the linearly polarized, slide incident, radiation, where α is the angle between the transmission direction of the P polarizer and the optic axis of the crystalline slide.

We analyze the phenomena which take place between the polarizer and the analyzer. Let

$$E_{in} = \begin{pmatrix} E_{x0} \\ 0 \end{pmatrix}$$

be the linearly polarized, on the direction of Ox, field of radiation which enters the slide, and

$$E_{em} = \begin{pmatrix} E_x \\ E_y \end{pmatrix}$$

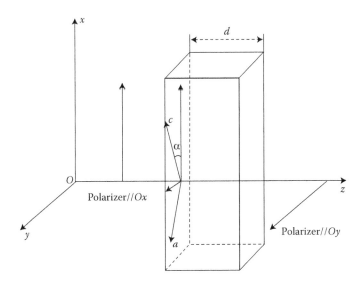

FIGURE 5.1 Propagation of light through a uniaxial, anisotropic medium, placed between two polarizers with perpendicular transmission directions.

the slide-emergent radiation field (Moisil and Moisil 1973, Malherbe 2007). The relationship between the electric field incident on the slide and the emerging electric field is described by relation (5.1).

$$\begin{pmatrix} E_x \\ E_y \end{pmatrix} = \begin{pmatrix} \cos\dfrac{\delta}{2} - i\sin\dfrac{\delta}{2}\cos 2\alpha & -i\sin\dfrac{\delta}{2}\sin 2\alpha \\ -i\sin\dfrac{\delta}{2}\sin 2\alpha & \cos\dfrac{\delta}{2} + i\sin\dfrac{\delta}{2}\cos 2\alpha \end{pmatrix} \begin{pmatrix} E_{x0} \\ 0 \end{pmatrix} \qquad (5.1)$$

Owing to the varied speeds of propagation, a dephasing $\delta = \delta_y - \delta_x$ occurs between the two waves, which determines the change of the polarization state of the radiation passing through the anisotropic substance layer

$$\delta = \frac{2\pi}{\lambda}\Delta n d \qquad (5.2)$$

We will consider the angle $\alpha = 45°$. Only the $E_{y,em}$ component, which has an oscillation direction parallel to the transmission direction of the analyzer, Oy, can determine the expression of the radiation intensity, emerging from the system ($I = E_{y,em} E^*_{y,em}$).

The exit intensity I is linked to the incident intensity I_0, as described in the relation

$$T(\lambda_0) = \frac{I}{I_0} = \sin^2\frac{\pi}{\lambda_0}\Delta n(\lambda_0)d \qquad (5.3)$$

where $T(\lambda_0)$ is the system transmission factor.

The representation of the system transmission factor, relative to the wavelength (wave number), is a transmission-channeled spectrum.

5.4 DETERMINATION OF QUARTZ BIREFRINGENCE FROM THE STUDY OF CHANNELED SPECTRA

The methods to measure the birefringence involve the determination of a dephasing δ, which allows the deduction of the optical path variation Δnd, for the considered wavelength, according to relation (5.2).

Interferometric methods can be used to obtain and analyze the channeled spectra—in the case of thick slides—or optical path compensation methods, which rely on using the *Babinet* compensator or a quarter-wave slide, when the optical path variation between the ordinary and the extraordinary wave is inferior to the wavelength.

Let the channeled spectrum (Figure 5.2), traced for a 1-mm-thick quartz crystal, with its facets cut parallel to the optic axis direction (*Oc*), be placed between two polarizers with perpendicular transmission directions. The neutral lines of the crystal are oriented at a 45° angle to the transmission directions of the polarizers.

Knowing the dependency of the wavelength transmission factor, it is possible to find the wavelengths of the radiations which maintain their polarization state after passing through the system (with the minimal values from Figure 5.2 corresponding to them) and also of those which maintain their polarization state but change their azimuth (with the maximal values from Figure 5.2 corresponding to them).

The increase in thickness of the anisotropic layer determines a rise in the number of maxima and null minima and facilitates the use of the desired spectral domain.

By eliminating, from the white light beam, the radiations with wavelengths corresponding to the transmission minima, a modification, in the spectral composition of the radiation emerging from the system, occurs.

The channeled spectrum of a nondispersive medium allows for the determination of the birefringence according to the relation

$$\Delta n = \frac{q\lambda_1\lambda_2}{d(\lambda_2 - \lambda_1)}$$ (5.4)

where q represents the number of minima between two λ_1 and λ_2 wavelengths.

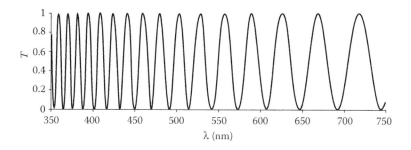

FIGURE 5.2 Transmission factor of a quartz slide between two crossing polarizers.

The values of quartz birefringence, determined by using this method, are presented in Table 5.1.

By using the "xyExtract Graph Digitizer" application from the LABfit curve-fitting software (http://zeus.df.ufcg.edu.br/labfit/download.htm), the wavelengths corresponding to the minima were extracted from the spectrum. Relation (5.4) for $q = 1$ (successive minima) was applied. Considering quartz to be a nondispersive medium, the average value of the birefringence (with a corresponding squared deviation) can be calculated: $\Delta n = 0.0108 \pm 0.0008$.

Another way of determining the birefringence of a nondispersive medium implies establishing the interference orders corresponding to the minima, and graphical representation of the quantity $1/\lambda = f(m)$, based on relation (5.5):

$$\frac{1}{\lambda} = \frac{m}{\Delta n d} \tag{5.5}$$

The m order of interference (Table 5.1) is determined (for the limit of the spectrum from large wavelengths) by considering the optical path variations of a neighboring minimum and maximum to be approximately equal.

The birefringence ($\Delta n = 0.0107$) can be determined from the slope of the graph (Figure 5.3).

TABLE 5.1

Orders of the Minima (m), the Number of the Minima (q), and the Birefringence (Δn) Determined from the Channeled Spectrum of a 1-mm-Thick Quartz Slide

λ (nm)	$1/\lambda$ (m^{-1})	m	q	Δn
353	2,832,861.19	28	17	0.0117
364	2,747,252.75	27	16	0.0124
375	2,666,666.67	26	15	0.0112
388	2,577,319.59	25	14	0.0120
401	2,493,765.59	24	13	0.0111
416	2,403,846.15	23	12	0.0112
432	2,314,814.81	22	11	0.0108
450	2,222,222.22	21	10	0.0111
469	2,132,196.16	20	9	0.0105
491	2,036,659.88	19	8	0.0105
515	1,941,747.57	18	7	0.0103
542	1,845,018.45	17	6	0.0103
572	1,748,251.75	16	5	0.0099
607	1,647,446.46	15	4	0.0100
646	1,547,987.62	14	3	0.0097
692	1,445,086.71	13	2	0.0097
745	1,342,281.88	12	1	

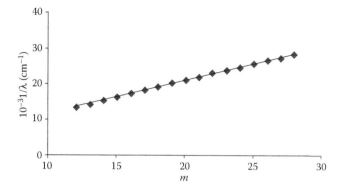

FIGURE 5.3 Dependency of the $1/\lambda$ quantity, relative to the m order of interference for the quartz crystal sample.

A variant of this method implies determination of the birefringence without ascertaining the order of interference. We number the minima and order them in an ascending order relative to the wavelength decrease and we represent the function $1/\lambda$ in relation with the values of q, where q is the number of the respective minimum.

From the slope of the graph, we determine the birefringence $\Delta n = 0.0108$.

We observe that the results obtained by using the three methods are comparable from a precision standpoint. However, those methods are not recommended for dispersive media, such as quartz.

Subsequently, we will describe a way of determining the birefringence which takes dispersion into account. From the condition for the minima (Macé De Lépinay 1885, 1892) (m is the order of interference for the minima), we obtain

$$\Delta n_{2m}(\lambda) = \frac{m\lambda_{2m}}{d} \tag{5.6}$$

The values of the birefringence, thus calculated, are described in Table 5.2.

One considers that the birefringence dispersion is described by Cauchy's relation (Medhat et al. 2003) (where $\tilde{v} = 1/\lambda$ is the wave number) $\Delta n = A + B\tilde{v}^2$.

Graphically representing the calculated values for the birefringence $\Delta n = f(\tilde{v}^2)$, the coefficients for Cauchy's dispersion relation can be determined from the equation of the straight line in Figure 5.4, $\Delta n = 8.6908 \times 10^{-3} + 1.5078 \times 10^{-16}\,\tilde{v}^2$.

The relation allows the determination of the birefringence regardless of wavelength. By comparing with the data from the literature (Cone 1990), we can observe that the precision of the determinations is quite good.

5.5 DETERMINATION OF THE BIREFRINGENCE OF A THIN SLIDE (AN ANISOTROPIC POLYMER FILM)

The birefringence phenomenon (double refraction) occurs in the case of certain transparent materials when subjected to mechanical stress. There is a relationship between the stress axes and the main axes of the materials subjected to the stress.

TABLE 5.2

Birefringence of the Quartz Slide

λ (nm)	m	Δn (Calculate)
353	28	0.0099
364	27	0.0098
375	26	0.0098
388	25	0.0097
401	24	0.0096
416	23	0.0096
432	22	0.0095
450	21	0.0095
469	20	0.0094
491	19	0.0093
515	18	0.0093
542	17	0.0092
572	16	0.0092
607	15	0.0091
646	14	0.0090
692	13	0.0090
745	12	0.0089

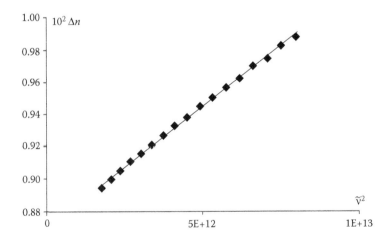

FIGURE 5.4 Representation of Cauchy's dispersion relation for the quartz crystal sample.

In the process of stretching the polymer films, a preferential orientation along the stretching direction of the macromolecules takes place. It is comparable to the optic axis of a uniaxial medium.

The dephasing δ (5.2), in the case of an anisotropic polymer film, occurs between the linearly polarized waves either parallel or perpendicular to the stretching direction of the film. These waves propagate through the anisotropic film with different speeds, determined by the refractive indexes. $\Delta n = (n_{//} - n_{\perp})$ represents the birefringence of the polymer film.

A thin polymer foil L_x introduces an optical path variation, $(\Delta) = (\Delta n \cdot d)$ comparable to the wavelength of the used radiation. When it is placed between two perpendicular polarizers, the resulting channeled spectrum has few minima and maxima (Pop et al. 1992–1993).

Next, we describe methods for determining the birefringence of thin films (Pop et al. 1994), using, in each case, the modifications brought by the transparent anisotropic film to the channeled spectrum of a thick anisotropic crystal L (Pop and Dorohoi 1991).

After obtaining the channeled spectrum of the L slide (Figure 5.5), the L_x film is introduced parallel to L and orientated with its stretching directions at a 45° angle to the transmission directions of the polarizers. The channeled spectrum that results in the presence of the thin slide is different from the previously obtained one. The changes depend on the birefringence and thickness of the film. In this system, the L slide is used to find the wavelength values for which the birefringence of the thin anisotropic slide is determined.

The channeled spectra for the orientation of the L_x layer are traced so that the high-speed directions of the two anisotropic media are parallel. We can observe a rise in the number of channels, as compared to the case in which the polarizers are separated solely by the L slide.

Between those spectra, the maxima are obtained for λ_{2m+1} wavelengths and the minima for λ_{2m} wavelengths.

The channeled spectrum for the case in which the low-speed direction of a slide is parallel to the high-speed direction of the other is also traced. In this channeled spectrum, the maxima are determined for λ_{2p+1} and the minima for λ_{2p} wavelengths. The values for m and p are determined from relations, such as $m = m_0 \pm q$; $p = p_0 \pm q$; $q = 0,1,2,\ldots$ in which m_0 and p_0 are integers and correspond to the first (or the last) minima (or maxima) from the obtained channeled spectrum.

The anisotropic polymer films were prepared using a 15% concentrated poly(vinyl alcohol) (PVA) solution obtained by dissolving PVA in water for approximately 5 h at 90°C. The resulting gel was filtered through a textile material and placed on clean

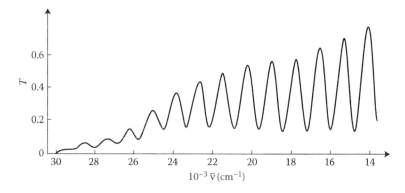

FIGURE 5.5 Channeled spectrum when the stretching direction of the L_x anisotropic film is perpendicular to the high-speed direction of the thick L slide.

TABLE 5.3

Orders of Interference and Their Corresponding Wavelengths for the (PVA) Films Sample

$\gamma = 2.41$			
λ (Å)	m	λ (Å)	p
6757	12	6902	9
6250	13	6197	10
5800	14	5636	11
5411	15	5161	12
5058	16	4762	13
4783	17	4431	14
4522	18	4125	15
4288	19	3877	16
4077	20	3652	17
3888	21	3488	18
3725	22	–	–
3571	23	–	–

glass panes, laying on a flat surface. Those were left to dry, at ambient temperature, without any external interventions to avoid alteration of their surface properties.

The stretching resistance of the films was studied from the mechanical properties point of view. The degree of stretching γ was evaluated from the ratio of the semi-axes of an ellipse in which a circle drawn on the dry polymer foil degenerates. Upon stretching, the polymer films have become uniaxial anisotropic media with the optic axis parallel to the stretching direction.

The channeled spectra were consequently traced in transmission factor (T) and wavenumber coordinates ($\tilde{v}$) for films with varying stretching degrees are described in Figure 5.5.

The orders of the minima and the corresponding wavelengths were extracted from the spectrums, as described in Table 5.3.

The (λ_{2m}, m) and (λ_{2p}, p) pairs of points were graphically represented as described in Figure 5.6, for a $\gamma = 2.41$ stretching degree.

The continuous variation of the birefringence, relative to the wavelength, allows the points from the graph in Figure 5.6 to be connected by a continuous curve. Working with the minima, the birefringence can be determined depending on their corresponding wavelengths.

$$\Delta n(\lambda) = \frac{m(\lambda) - p(\lambda)}{2h} \lambda \qquad (5.7)$$

Through interpolation, the birefringence can be determined, not only for the wavelengths corresponding to the minima or maxima but also for any wavelength from the visible spectrum. For the λ wavelength from Figure 5.6, the $m(\lambda) - p(\lambda)$

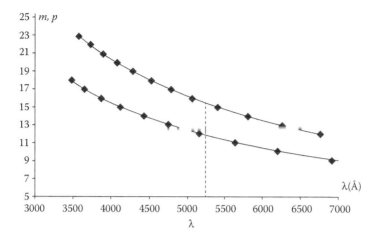

FIGURE 5.6 Order of interference variation relative to the wavelength.

variation, as determined with the help of the continuous curves, allows the calculation of the $\Delta n(\lambda)$ birefringence based on relation (5.7).

5.6 APPLICATIONS OF THE BIREFRINGENCE INDUCED ON THE ANISOTROPIC POLYMER FILMS

The development of compensatory slides can be estimated by the polarization state of the radiations after their passage through an anisotropic PVA film.

5.6.1 INDUCED BIREFRINGENCE

Based on their supramolecular structure, macromolecular compounds can be isotropic, when the polymer chains have a random orientation, or crystalline, if the macromolecular chains (having large lengths when compared to their thicknesses) are packed on variable distances with the formation of some areas with a high crystallographic orientation, which can be highlighted by using x-rays. A polymer material becomes birefringent when its chains are orientated, as a consequence of the intrinsic optical anisotropy (Doi and Edwards 1994). This type of birefringence is called orientation birefringence. If the polymer films are subjected to stretching, an additional induced birefringence is added to the orientation birefringence.

The main refractive indices (ordinary and extraordinary) of the polymer film, as measured in linearly polarized light, with an electric field intensity oscillation direction orientated both parallel and perpendicular to the stretching direction, are usually different.

The birefringence of the stretched polymer films could be considered a measurement of the orderliness degree of the polymer chains in the PVA sample.

The induced birefringence dependence on the stretching degree for various thicknesses of the PVA films (Nechifor et al. 2010) is described in Figure 5.7.

The induced birefringence depends on the stretching degree for various thicknesses of the PVA films. The birefringence tends to decrease with the increase in

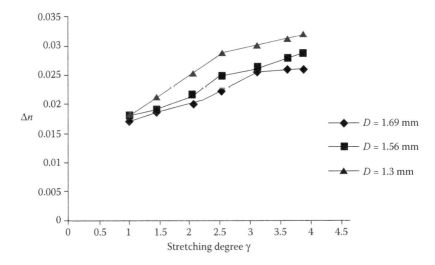

FIGURE 5.7 Birefringence versus the stretching degree for various thicknesses of the PVA films.

thickness of the film. For a given thickness, the birefringence tends to increase linearly with the stretching degree, for degrees smaller than 2.5. For stretching degrees larger than that, the dependency remains linear but has a smaller slope.

5.6.2 POLARIZATION STATE

The polarization state is conventionally described by the electric field intensity vector. For a plane monochromatic wave which propagates on the z axis direction (Kužel 2000/2001), the electric field vector is in the xy plane and it is decomposed into two components: E_x and E_y. The two components satisfy the equation of an ellipse, thus, the elliptical polarization corresponds to the light's general state of polarization. The shape of the ellipse depends both on the dephasing between the two individual oscillations and the size of the two amplitudes

5.6.3 DEVELOPMENT OF THE COMPENSATORY SLIDES

We analyzed the possibility of using PVA films for making the compensatory slides (full, semi, and quarter-wave slides) (Angheluţă 2011). We calculated the change in the state of polarization for a monochromatic radiation after passing through the anisotropic polymer layer. For a 1.3-mm-thick polymer film, whose stretching degree varies between 1.00 and 4.00, the birefringence versus stretching degree is shown in Figure 5.7.

We consider the curve as being composed of two linear sections.

The change of polarization state of the monochromatic radiation with the wavelength of 589.3 nm will be analyzed. The polarization state of the emerging radiation was estimated for various degrees of stretching. Once the stretching degree is modified, we have a change in the birefringence and the phase shift δ.

TABLE 5.4
Polarization State of Radiation Depending on the Stretching Degree of a 1.3-mm-Thick Polymer Film Sample

γ	Δn	δ (degree)	T	Polarization State	φ (W/m²)	
2.46	0.0282	91.27	0.5111	Right circular	102.21	Quarter-wave slides
2.67	0.0288	224.44	0.8570	Left elliptical	171.40	–
3.04	0.0297	180.27	1.0000	Linearly with changed azimuth	200.00	Semiwave slides
3.36	0.0304	44.78	0.1451	Right elliptical	29.02	–
3.73	0.0313	0.61	0.0000	Linear	0.01	Full-wave slides

In order to illustrate the variations of the polarization state, as determined by the anisotropic film, we developed a Microsoft Excel application which, among others, allows to determine

The system's transmission factor, as described in relation (5.3).
The intensity values (the energetic flux density φ) for the radiation emerging from the polarizer, film, and analyzer system.

The polarization ellipse can be obtained by composing the radiations of the electric field intensity vectors which have perpendicular action directions. The initial phases of the two components when entering the slide are equal (they originate from a linearly polarized radiation). After propagating through the slide, the components are dephased by δ. By knowing that the birefringence is relative to the stretching degree, and the wavelengths, it is also found that δ dephasing varies with the modification of the stretching degree. We considered the value of the dephasing, introduced with the propagation on the Ox axis, as being the origin of the dephasing. The net phase difference is equal to the one introduced with the propagation on the Oy axis. This determines a modification of shape for the polarization ellipse.

The energetic flux density of the incident radiation was considered to be 200 W/m².

We selected the stretching degree values which allowed us to obtain the desired states of polarization (those that were useful from a practical point of view), and described them in Table 5.4. Owing to the periodicity of the harmonic functions that intervene in the described relations, the dephasing values were reduced to the 0–2π interval.

5.7 CONCLUSIONS

The foil acts as a wavelength blade (γ = 3.73) if the radiation remains linearly polarized, with the same azimuth after passing through the anisotropic film. In this situation, the flux density and the postanalyzer transmittance are minimal. When the radiation is linearly polarized with a changed azimuth, the foil acts as a semi-wavelength blade (γ = 3.04); so, in this case, the flux density and the transmittance are maximum. The oscillating direction of electric field intensity of the emerging radiation is parallel to the transmission direction of the analyzer. For a phase shift $(2m + 1)\pi/2$, the foil acts as a quarter-wavelength blade (γ = 2.46).

The accuracy of determining the degree of stretching allows us to find a good approximation of the phase shift values corresponding to cases of polarization states, close to the ideal case.

REFERENCES

Angheluță, E. 2011. Determining the polarization state of the radiation crossing through an anisotropic poly (vinyl alcohol) film. *Rom. J. Phys. Buchar.* 56(7–8):971–975.

Chirica, L., Dorohoi, D., Pop, V., and Strat, M. 1985. Determination of birefringence and of main indices of refraction of polymer foils Analelele ştiinţifice ale Universităţii "Al.I.Cuza" Iaşi, T XXXI, 53. Iaşi

Cone, G. 1990. *Optica electromagnetică a mediilor anizotrope [The electromagnetic optics of anisotropic mediums]*. Editura Tehnică, Bucureşti, 24, 25.

Delibas, M. and Dorohoi, D.O. 1999. *Indrumar de Optică [Guide to Physical Optics]*. Litografia Univ. Al.I.Cuza, Iaşi.

Doi, M. and Edwards, S.F. 1994. *The Theory of Polymer Dynamics*. Clarendon Press, Oxford, 122.

Dumitraşcu, L., Dumitraşcu, I., and Dorohoi, D. 2006. *Complemente de fizică pentru studenţii şcolilor doctorale [Complementary physics material for graduate school students]*, vol. I, 111–114. Tehnopress, Iaşi.

Giancoli, D.C. 2004. *Physique générale: Ondes, optique et physique modern [General physics: Waves, optics and modern physics]*. Montreal, 183.

Halliday, D. and Resnick, R. 1975. *Fizica*, vol. II [Physics vol. II]. Editura Didactică şi Pedagogică, Bucureşti, 519.

Kužel, P. 2000/2001. *Electromagnétisme des milieux continus—OPTIQUE—Licence de Physique* [Continuum electromagnetism—Optics]. Institut Galilée, Université Paris-Nord, 31–38.

Mace de Lepinay, M.J. 1885. Etude de la dispersion de double refraction du quartz [Studying the dispersion of double refraction of quartz]. *J. Phys. Theor. Appl.* 4(1):159–166. <10.1051/jphystap:018850040015901>. <jpa-00238323>

Mace De Lepinay, M.J. 1892. On the double refraction of quartz. *J. Phys. Theor. Appl.* 1(1):23–31. <10.1051/jphystap:01892001002300>. <jpa-00239611>

Malherbe, J.M. 2007. Principes theoriques du filtre monochromatique de Solc (Aout) [Theoretical principles of the Monochromatic solc filter]. http://solaire.obspm.fr/images/documentation/Filtre-SOLC.pdf

Medhat, M., Hendawy, N.I., and Zaki, A.A. 2003. Interferometric method to determine the birefringence for an anisotropic material. *Egypt J. Sol.* 26(2):231–238, http://egmrs.powweb.com/EJS/PDF/vo262/231.pdf

Moisil, D. and Moisil, G. 1973. *Teoria si practica elipsometriei [Theoretical and practical aspects of ellipsometry]*. Editura Tehnică, Bucureşti, 18–22, 29.

Nechifor, C.-D., Angheluta, E., and Dorohoi, D.-O. 2010. Birefringence of retired poly (vinyl alcohol) (PVA) foils. *Materiale Plastice*, 47(2):164.

Pop, V. 1988. *Bazele opticii [The basics of optics]*. Editura Universitatii "Al.I Cuza," Iaşi.

Pop, V. and Dorohoi, D. 1991. Brevet Romania 10103972/27.06.1991 [Romania Patent 10103972/27.06.1991].

Pop, V., Dorohoi, D.O., and Crîngeanu, E. 1994. A new method for determining birefringence dispersion. *J. Macromol. Sci. Phys.* B33(3/4):373.

Pop, V., Strat, M., and Dorohoi, D. 1987. Brevet Romania 93528/31-08-1987 [Romania Patent nr. 93528/31-08-1987].

Pop, V., Dorohoi, D., and Cringeanu, E. 1992–1993. La determination de la dispersion de la birefringence aux feuilles minces [Determining the dispersion of the birefringence in thin films]. *Analele Ştiint.* Univer. "Al.I.Cuza," Iaşi, 457–462.

6 New Approaches on Birefringence Dispersion of Small-Molecule Liquid Crystals Designed as Interferential Optical Filters

Irina Dumitrascu, Leonas Dumitrascu, and Dana Ortansa Dorohoi

CONTENTS

6.1 INTRODUCTION

Liquid crystals (LCs) constitute a class of substances, which, under certain thermodynamic conditions, present both the properties of fluids (flowing, forming drops) and solids, such as anisotropy, ordering of the composing particles (Demus et al. 1998).

The stable thermodynamic states through which the substances pass during the modification of the thermodynamic parameters are called mesophases. The substances that can exist in LC state are called mesogene substances, and their molecules are called mesogenic molecules (Dumitraşcu and Dorohoi 2015).

The LCs classification criteria are established depending on the presence of the following characteristics: the existence of a translation ordering, the existence of an orientation ordering of the chemical bonds, the existence of a correlation between the layers in case of the smectic LCs, the presence of chirality, and the presence of a cubic structure (Muscutariu 1981).

Another choice of the LCs classification criteria holds in view elements, such as (Demus et al. 1998)

- The ways of obtaining and the parameter controlling the LC state
- The structural aspect revealed by the analysis with a polarizing microscope
- The structure and shape of the molecules and constituent molecular blocks

1. Depending on the obtaining method and the parameter controlling their state, LCs can be classified into two groups: thermotropic LCs and lyotropic LCs (Dumitraşcu and Dorohoi 2015).

   Thermotropic LCs are obtained by melting solid crystals, while the lyotropic ones are obtained by solving mesogenic molecules—usually amphiphilic—into polar solvents (most of the time water) or nonpolar solvents (carbon tetrachlorure).

   The lyotropic LC molecules are greater than the thermotropic LC molecules. The axis ratio of lyotropic LCs is 15 to 4–8 in thermotropic LCs. The molecular mass of thermotropic LC is 200–500 g/mole and each molecule participates equally to the realization of its ordering. From a structural point of view, the lyotropic LCs form only smectic structures (Dumitraşcu and Dorohoi 2015).

   The parameter controlling the thermotropic LC state is temperature, while the one that conditions the obtaining of lyotropic LCs is concentration (Demus et al. 1998).

2. Considering the structural aspect revealed by the polarizing microscope, in 1922, G. Friedel proposed the first scheme for the classification of thermotropic LCs into the following three groups (Dumitraşcu and Dorohoi 2015):
   - Nematic LCs (from the Greek nematos which means thread)
   - Cholesteric LCs (also called chiral nematic)
   - Smectic LCs (from the Greek smectos which means soap)

   The main cause of the structural aspect revealed by the polarizing microscope analysis of the thermotropic LCs consists of the spatial distribution and orientation of the molecules, characteristic to each mesogenic phase.

3. According to the shape of the molecules or the molecular blocks composing the mesogenic molecules, the LCs can be grouped into several categories: calamitic LCs, discotic LCs, and polycatenar LCs (Demus et al. 1998).

   The molecules that are preferentially oriented on a direction, which involves an advanced degree of orientational order, but a low degree of positional ordering, can be organized in nematic phase. It is a low-temperature phase, which in relatively high temperatures goes through an independent transition toward the isotropic liquid phase. The elongated sensitive form of the molecules or the molecular groupings constitutes a sine qua non condition of mesomorphic phases. In the nematic LC, the molecules can only move along their long axes, from side to side and from top to bottom.

The preferential direction of the long axes of the molecules can be easily modified by external factors, such as the electric field, the magnetic field, temperature, and the contact of the molecules with the walls of the recipient containing the LC (Demus et al. 1998).

It was noticed from the optical analysis of the nematic mesophases that an ideal balance configuration is indicated by a transparency close to 100%. An organic molecule that is typical for the nematophase is that of the *p*-azoxyanisol (PAA), which has the cross-section diameter approximately four times smaller than the length of its main axis. Also, nematogene, but more accessible, is the substance called *N*-(*p*-methoxybenzylidene)-*p*-butylaniline (MBBA). The nematic phase is carried out with MBBA within the interval 20–40°C (Diaconu et al. 2007).

Inside the nematic LCs, the molecules are ordered with the long axes arranged parallel, and the mass centers of the molecules distributed chaotically. The parallel arrangement of the long axes of the molecules and the creation of a parallel director field can be obtained either when the thermal fluctuations and all the exterior interactions are negligible, or in an electric field, a magnetic field of certain intensities, or due to the interactions with the surfaces of the precinct walls (Dumitraşcu and Dorohoi 2015).

6.2   ANISOTROPY OF THE ELECTRO-OPTICAL PROPERTIES OF THE LCs

Anisotropy is a substance characteristic which involves not having the same physical properties in all directions and it depends on the molecular structure and the way in which the crystalline network is formed (Bruhat 1965, Pop 1988). Owing to the preferential ordering of the molecules, the LCs are generally anisotropic media whose properties can only be described by tensorial parameters (Pop 1988, Born and Wolf 1999).

In the LCs formed of polar molecules, the induced polarization (electronic and ionic) is added to the orientation polarization due to the tendency of the permanent electric dipole moments to orient under the action of the external electric field, parallel to its lines (Dumitraşcu and Dorohoi 2015).

The optical anisotropy of the LCs determines the propagation of the polarized light along the propagation direction with a different speed compared to the propagation speed on the directions perpendicular to it. Consequently, the LCs are birefringent media (Demus et al. 1998).

The optical axis of an LC is defined as the propagation direction of the optical radiations, for which the propagation speed is independent of their polarization state. The optical axis is generally parallel to the symmetry axis of the highest order (Scharf 2006).

The uniaxial LCs have a single optical axis and two different values for the main refraction indices (Scharf 2006):

- The ordinary refraction index n_o corresponds to the linearly polarized light, whose vibration vector (electric component) is oriented perpendicularly to the optical axis (ordinary ray).

- The extraordinary refraction index n_e corresponds to the linearly polarized light, whose vibration vector (electric component) defines, together with the propagation direction, a plane containing the optical axis (extraordinary ray).

Birefringence is the physical parameter characterizing the totally or partially ordered anisotropic media; in the case of uniaxial LCs, it is expressed by the relation $\Delta n = n_e - n_o$ (Savelyev 1980, Pop 1988). If $\Delta n > 0$, the medium is called uniaxial positive and if $\Delta n < 0$, the medium is called uniaxial negative (Pop 1988).

6.3 EXPERIMENTAL DETERMINATION OF OPTICAL BIREFRINGENCE FOR MBBA LC

For the study conducted in the visible range, the liquid crystalline layer made from MBBA was kept in a special cell (Figure 6.1) consisting of two glass walls provided with interior conducting layers of SnO_2 deposed on their internal sides in order to permit the application of an external electrostatic field, perpendicularly oriented on the glass plate's surfaces (Diaconu et al. 2007).

The parallelism of the two glass plates is ensured with the help of four spacers that also determine the thickness of the LC layer at a value $L = 14$ μm. The cell walls were cleaned in distilled water using an ultrasonic method and the conducting layer was covered with a special thin molecular orientation layer of lecithin, obtained by slowly passing the internal walls of the cell through a solution of 5% lecithin in water. This layer determines the orientation of the nematic LC molecules with the axes in a plane parallel to the facets of the two glass plates, so that the optical axis of the nematic LC is perpendicular to the light propagation direction (Dorohoi et al. 2001).

After the cell was filled with LC, its walls were sealed by applying a synthesis polymer (epoxy resin) on the external contour. The dominant orientation of the molecular axes along the walls was also mechanically ensured.

The main values of the refractive index were measured with linearly polarized waves. The ordinary value (n_o) was measured with light having the electric field

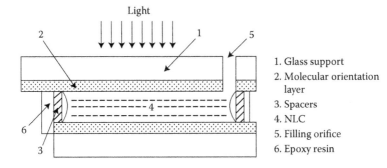

FIGURE 6.1 The cell where the liquid crystal was located.

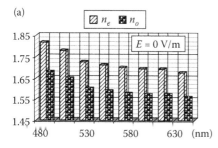 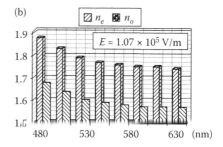

FIGURE 6.2   Main refractive indices of MBBA liquid crystal versus the wave length in the visible range in the absence (a) and in the presence (b) of an electrostatic field.

perpendicular to the optical axis and the extraordinary value (n_e) was measured with linearly polarized light having its electric field intensity parallel to the optical axis (Picos et al. 2005, Dumitraşcu et al. 2007b).

The measurements were made in the absence and in the presence of the external electrostatic field applied between the opposite walls of the cell and having a variable intensity (Dumitraşcu et al. 2007b).

Some obtained results regarding the refractive indices of MBBA LC are given in Figure 6.2a and b (Dumitraşcu et al. 2007a).

From the graphics presented in Figure 6.2, one can see that

- The values of the main refractive indices and of the corresponding birefringence depend on the wavelength of the radiation for which it was determined. (Both the principal refractive indices and the birefringence of the studied samples decrease with the light wavelength increasing. This means that MBBA shows a normal dispersion.)
- These values are influenced by the external electrostatic fields applied to the studied LC layers. (The main refractive indices and the corresponding birefringence increase with the increase in the intensity of the external electrostatic field.)
- The ordinary refractive indices of MBBA do not depend considerably on the intensity of the applied external electrostatic field, while the applied external electrostatic field has an important influence on the values of the extraordinary refractive indices.
- In the case of MBBA, a saturation tendency can be noted for the extraordinary refractive index and for birefringence near the value of the electrostatic field intensity produced by the voltage of 2 V applied to the LC cell.

6.4   MODELING AND SIMULATING THE DISPERSION OF BIREFRINGENCE

In general, the refractive index of a transparent medium is a function of the frequency v of the light; thus, $n = n(v)$, or alternately, with respect to the wave's wavelength $n = n(\lambda)$.

The wavelength dependency of a material's refractive index is usually quantified by some empirical formula, such as the Cauchy equation, the Hartman formula, or the Sellmeier equation (Dorohoi et al. 2006).

The Cauchy equation is an empirical relationship between refractive index n and wavelength λ for a particular transparent medium; it was proposed for modeling the dispersion phenomenon.

Using the Cauchy formula, the ordinary and the extraordinary refractive indices dispersion can be expressed by the following relations (Dumitraşcu et al. 2006a):

$$n_{o,e}(\lambda) = A_{o,e} + \frac{B_{o,e}}{\lambda^2} + \frac{C_{o,e}}{\lambda^4} \tag{6.1}$$

where A_i, B_i, and C_i ($i = e,o$) must be determined from experimental data.

In order to determine all the Cauchy coefficients, it is necessary to know the values of the refractive indices for three different wavelengths and to solve the following system of equations (Dumitraşcu et al. 2007):

$$\begin{cases} n_1^2 = A + \dfrac{B}{\lambda_1^2} + \dfrac{C}{\lambda_1^4} \\[2mm] n_2^2 = A + \dfrac{B}{\lambda_2^2} + \dfrac{C}{\lambda_2^4} \\[2mm] n_3^2 = A + \dfrac{B}{\lambda_3^2} + \dfrac{C}{\lambda_3^4} \end{cases} \tag{6.2}$$

The solution of system (6.2) is

$$\begin{cases} A = \dfrac{\lambda_1^4 n_1^2\left(\lambda_3^2 - \lambda_2^2\right) + \lambda_2^4 n_2^2\left(\lambda_1^2 - \lambda_3^2\right) + \lambda_3^4 n_3^2\left(\lambda_2^2 - \lambda_1^2\right)}{\lambda_1^4\left(\lambda_3^2 - \lambda_2^2\right) + \lambda_2^4\left(\lambda_1^2 - \lambda_3^2\right) + \lambda_3^4\left(\lambda_2^2 - \lambda_1^2\right)} \\[4mm] B = \dfrac{\lambda_1^4 \lambda_2^4\left(n_1^2 - n_2^2\right) + \lambda_3^4 \lambda_1^4\left(n_3^2 - n_1^2\right) + \lambda_2^4 \lambda_3^4\left(n_2^2 - n_3^2\right)}{\lambda_1^4\left(\lambda_3^2 - \lambda_2^2\right) + \lambda_2^4\left(\lambda_1^2 - \lambda_3^2\right) + \lambda_3^4\left(\lambda_2^2 - \lambda_1^2\right)} \\[4mm] C = \dfrac{\lambda_1^4 \lambda_2^2 \lambda_3^2 n_1^2\left(\lambda_3^2 - \lambda_2^2\right) + \lambda_1^2 \lambda_2^4 \lambda_3^2 n_2^2\left(\lambda_1^2 - \lambda_3^2\right) + \lambda_1^2 \lambda_2^2 \lambda_3^4 n_3^2\left(\lambda_2^2 - \lambda_1^2\right)}{\lambda_1^4\left(\lambda_3^2 - \lambda_2^2\right) + \lambda_2^4\left(\lambda_1^2 - \lambda_3^2\right) + \lambda_3^4\left(\lambda_2^2 - \lambda_1^2\right)} \end{cases} \tag{6.3}$$

The fitting coefficients A_i, B_i, and C_i ($i = e,o$) of MBBA depend on the electric field intensity (Dumitraşcu et al. 2007).

In the linear approximation, they can be calculated with the relations

$$\begin{cases} A_i(E) = A_i(0) + \alpha_i \cdot E \\ B_i(E) = B_i(0) + \beta_i \cdot E \quad (i = e,o) \\ C_i(E) = C_i(0) + \gamma_i \cdot E \end{cases} \tag{6.4}$$

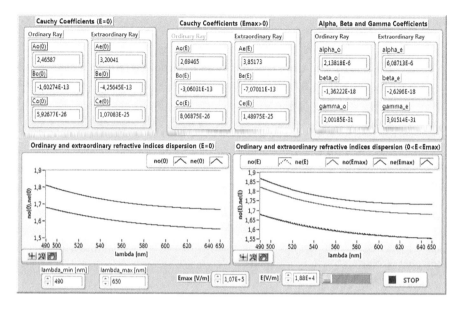

FIGURE 6.3 LabView application frame for computing the Cauchy fitting coefficients and for simulating of the ordinary and the extraordinary refractive indices dispersion.

The coefficients α_i, β_i, and γ_i ($i = e,o$) can be computed from the experimental data.

For computing all the Cauchy-fitting coefficients which appears in Equations 6.1 through 6.4 and for computer simulation of the ordinary and the extraordinary refractive indices dispersion, a LabView application was implemented; a screen capture of this application frame is shown in Figure 6.3.

The Cauchy-fitting coefficients for MBBA LC were calculated on the basis of the experimental data shown in Figure 6.2 (Dumitraşcu et al. 2007a,b), by using relations (6.1) through (6.4).

The values of the ordinary and extraordinary refractive indices used as input data for the LabView application shown in Figure 6.3 are listed in Tables 6.1 and 6.2.

TABLE 6.1

Ordinary and Extraordinary Refractive Indices of MBBA in the Absence of External Electric Field ($E = 0$)

No.	Lambda (nm)	n_o	n_e
1	490	1.6812	1.8185
2	560	1.5992	1.7123
3	623	1.5608	1.6751

TABLE 6.2
Ordinary and Extraordinary Refractive Indices
of MBBA in the Presence of an External Electric
Field ($F = 1.07 \cdot 10^5$ V/m)

No.	Lambda (nm)	n_o	n_e
1	490	1.6752	1.8775
2	560	1.5835	1.7641
3	623	1.5602	1.7365

The Cauchy-fitting coefficients A_i, B_i, and C_i ($i = e,o$), and the corresponding coefficients α_i, β_i, and γ_i calculated and displayed by the LabView application shown in Figure 6.3, are important because these values are used as input data for the simulations of transmission factor of the tunable polarization interferential filters presented in the next section.

6.5 LIQUID CRYSTAL POLARIZATION INTERFERENTIAL FILTERS MADE FROM MBBA

Tunable polarization interferential filters made of liquid crystals (LCTF) are increasingly frequent in applications from various fields, such as optical fiber communications, astronomy, pollution monitoring, color generation in displays, and medical diagnosis (Xia et al. 2002).

The large aperture and the wide spectral field provided by LC polarization interferential filters represent a particular advantage in what concerns the conventional dispersive spectral analysis techniques.

Some of the advantages of LC-tunable filters compared to acousto-optical tunable filters are low energy consumption, low source voltage, excellent image quality, and high clarity, including in the case of a wide aperture.

Interferential polarization filters include elements under the form of cells filled with LC, with electrically controlled birefringence, for the selection and transmission, at the exit, of a certain wavelength from the analyzed spectral field, the rest of the wavelengths being excluded. This type of filters is perfect in case of electronic devices designed to capture color images, such as digital cameras with CCD-type sensor (charge-coupled devices), as they provide excellent image quality, having an almost linear feature of the optical path difference compared to the applied voltage.

According to the control pattern used in order to select the transmitted wavelength, there are two distinct types of polarization interferential filters made of LCs: discrete and continuous.

The two types are based on the linearly polarized wave interference propagated through the filter elements, except that

- In case of continuous control tunable filters, the source voltage applied on all elements of the filter is continuously varied, in order to obtain a continuous modification of the wavelength selected by the filter.

- In case of discrete control filters, one uses a voltage that varies by leaps, which generates, at the filter exit, two complementary spectra.

Using several elements mounted in a cascade, one obtains increasingly narrow transmission bands, but this leads to a decrease in the transmission intensity and an increase in the filter complexity.

Polarization interferential filters with continuous control are fit for applications that do not need high resolutions.

6.5.1 Structure of a Liquid Crystalline-Tunable Filter Element

The basic structure of elements of the tunable filters polarization interferential filters made of LCs is shown in Figure 6.4a and b (Dumitraşcu et al. 2007a).

In order to avoid the use of very thick anisotropic layers and in order to perform a rigorous control of the optical features of the filter, it is recommended to obtain tunable elements made of two layers of LC.

6.5.2 Transmission Factor Computation

The two layers (one of which having electrically controlled birefringence) may have (Dumitraşcu et al. 2008):

Case I. Parallel optical axes, as in Figure 6.5a and b
Case II. Perpendicularly optical axes, as in Figure 6.6a and b

The transmission factor of light through the device (T) is dependent on the value of phase delay $\Delta\varphi$. The phase delay for the device illustrated in Figures 6.5 and 6.6 can be calculated with the relation (Holmes 1964, Dumitraşcu et al. 2006b,)

$$\Delta\varphi = \frac{2\pi(L_1 \cdot \Delta n_1 \cdot (\lambda, E) \pm L_2 \cdot \Delta n_2(\lambda))}{\lambda} \qquad (6.5)$$

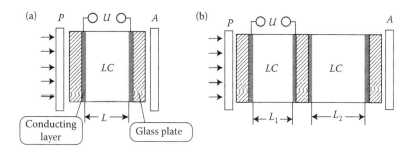

FIGURE 6.4 Schematic structure of a liquid crystalline-tunable filter element: (a) a single-layered LCTF element and (b) a double-layered LCTF element.

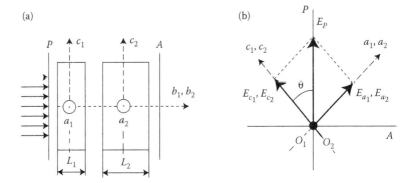

FIGURE 6.5   (a) The slow axes of the two layers are parallel; (b) the relative orientation of the polarizer transmission directions and of the basic directions of anisotropic media.

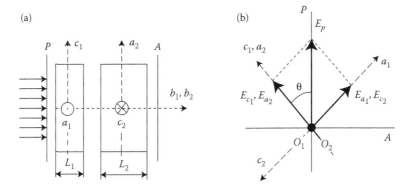

FIGURE 6.6   (a) The slow axes of the two layers are perpendicular; (b) the mutual orientation of the polarizer transmission directions and of the basic directions of anisotropic media.

The transmission factor of light through the device is given by one of the following relations (Dumitraşcu et al. 2008):

- For perpendicularly transmission directions of the two polarizers:

$$T = \frac{1}{2}\cos^2\left(\frac{\Delta\varphi}{2}\right) = \frac{1}{2}\cos^2\left(\frac{\pi(L_1 \cdot \Delta n_1(\lambda, E) \pm L_2 \cdot \Delta n_2(\lambda))}{\lambda}\right) \qquad (6.6)$$

- For parallel transmission directions of the two polarizers:

$$T = \frac{1}{2}\sin^2\left(\frac{\Delta\varphi}{2}\right) = \frac{1}{2}\sin^2\left(\frac{\pi(L_1 \cdot \Delta n_1(\lambda, E) \pm L_2 \cdot \Delta n_2(\lambda))}{\lambda}\right) \qquad (6.7)$$

Equations 6.6 and 6.7 show the transmission factor of a single-stage polarization interference filter (PIF), which is an oscillatory function of the path-length difference.

6.5.3 WOOD LC–PIF TRANSMISSION FACTOR

The basic structure of the tunable Wood polarization interferential filters made of LCs is shown in Figure 6.7a and b.

An important disadvantage of a Wood interferential filter with a single element is the equality of the bandwidth of the bands corresponding to the transmission maxima and the width of the channel.

Consequently, the Wood interferential filter with a single stage has a modest selectivity and its use in defining and transmission of some spectral channels of reduced width (e.g., in data communication by optical fibers) is nonperforming (Dumitraşcu et al. 2008).

In order to improve their performances, a multilayer geometry is adopted here. This kind of filters is achieved by more identical layers (Figure 6.7b).

We consider a device achieved from m identical elements so that the transmission direction of the entrance polarizer in each element is parallel to the transmission direction of the exit polarizer from the precedent layer. Then, one from the above-mentioned polarizer can be eliminated.

The transmission factor of the device becomes (Dumitraşcu et al. 2008)

$$T = \frac{1}{2}\sin^{2m}\left(\frac{\pi\left(L_1\Delta n_1(\lambda,E)\pm L_2\Delta n_2(\lambda)\right)}{\lambda}\right) \quad (m=2,3,4,\ldots) \tag{6.8}$$

The wavelengths corresponding to the maxims and to the minims of transmission for a Wood interferential filter with a single stage are

$$\begin{cases} \lambda_{k,\max} = \dfrac{2(L_1\Delta n_1 \pm L_2\Delta n_2)}{4k\pm 1}, & (k\in Z) \\[3mm] \lambda_{k,\min} = \dfrac{L_1\Delta n_1 \pm L_2\Delta n_2}{k}, & (k\in Z, k\neq 0) \end{cases} \tag{6.9}$$

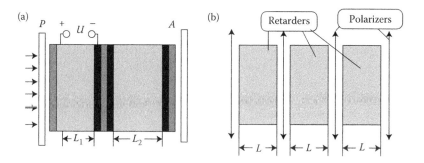

FIGURE 6.7   (a) Single-stage LCTF system and (b) three stages Wood polarizing interference filter.

The spectral bandwidth corresponding to the maxim from the wavelength $\lambda_{k,max}$ is

$$\Delta\lambda = |\lambda_{k,max} - \lambda_{k,min}| = \frac{\lambda_{k,max}}{k} \cong \frac{\lambda_{k,max}^2}{2(L_1\Delta n_1 \pm L_2\Delta n_2)} \qquad (6.10)$$

Relation (6.10) permits the estimation of the retardation $(L_1\Delta n_1 \pm L_2\Delta n_2)$ necessary to eliminate one from the two neighboring lines separated by the spectral interval $\Delta\lambda$.

The number of the transmitted bands in the spectral range limited by the wavelengths λ_1 and λ_2 $(\lambda_1 < \lambda_2)$ can be estimated by the relation

$$N \cong \frac{\lambda_2 - \lambda_1}{\lambda_1\lambda_2}(L_1\Delta n_1 \pm L_2\Delta n_2) \qquad (6.11)$$

The wavelengths corresponding to the maxims and minims of transmission and the number of the transmitted bands in the considered spectral range λ_1 and λ_2 $(\lambda_1 < \lambda_2)$ are the same with those obtained with a filter achieved by one element (relations (6.6) through (6.8) and (6.11), respectively).

The transmission spectral bandwidth neighboring the maxim of k order is modified (one decreases). It is given by the relation (Dumitraşcu et al. 2008)

$$\Delta\lambda_{(m)} = \frac{4\pi\left(L_1\Delta n_1(\lambda,E) \pm L_2\Delta n_2(\lambda)\right)\arcsin\dfrac{1}{2^m}}{(2k+1)^2\pi^2 - 4\left(\arcsin\dfrac{1}{2^m}\right)^2} \qquad (6.12)$$

Relations (6.8) through (6.12) are the equations of the model used for simulating Wood-tunable LC–PIFs shown in Figure 6.8.

The simulation window displays two graphics showing the transmittances versus the wavelengths corresponding to the two cases (Case I and Case II explained in Figure 6.5a and b—parallel optical axes and, respectively in Figure 6.6a and b—perpendicularly optical axes).

From Figure 6.8, one can see a clear difference in the half-pass band values which correspond to the two different orientations of the optical axis (in Case I, the half-pass band, around a wavelength of about 540 nm, is smaller than in Case II; this remark suggests that the orientation, corresponding to parallel optical axes of the two LC layer components of the basic stage, is more efficient in filtering the optical radiations around the wavelength of 540 nm).

From Figure 6.8, one can also see the increasing filtering effect produced by the increased number of stages. From this point of view, the better results, regarding half-pass band values, correspond to the five stages of Wood filter (TW-I_5 and TW-II_5).

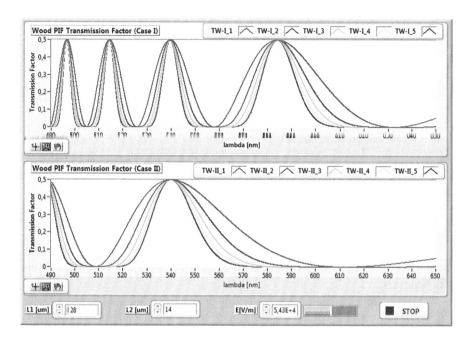

FIGURE 6.8   LabView application frame that simulates a single-stage and some multistage Wood PIF transmission factor.

6.5.4 LYOT–OHMANN LC–PIF TRANSMISSION FACTOR

Interferential polarization filters are typically grouped into two main categories: Solc structure filters (Solc 1965) respectively, Lyot structure filters (Lyot 1933, 1944). These two filter types have certain advantages and certain disadvantages; thus, a third filter type, the Evans type, is used, as a compromise (Dumitraşcu et al. 2008).

The structures taken into account here are based on the works of Lyot and Ohmann (Ohman 1938, 1958). These filters are made of a cascade of united filters or stages, each stage needing two polarizers (the transmission directions are parallel or perpendicular to Figures 6.5 and 6.6), linking several linear-phase changers, oriented at 45°, made of combinations of one or two LC layers (Figure 6.4a and b).

LC layers in a typical Lyot filter are often selected for a binary sequence of retardation so that the transmission value is maxim at the wavelength determined by the thickest crystal retarder (Figure 6.9b). Other stages in the filter serve to block the transmission of unwanted wavelengths.

By cascading a series of these filter stages, a band-pass filter can be synthesized. For the computational modeling presented below, the retardation values were chosen in binary steps of the crystalline thickness layers: (L_1, L_2), $(2L_1, 2L_2)$, and $(4L_1, 4L_2)$.

The global transmission factor of light through the three stages of Lyot-polarizing interference filter is given by the following equation (Dumitraşcu et al. 2008):

$$T = \frac{1}{2}\cos^2\left(\frac{\Delta\varphi}{2}\right)\cos^2\left(2\cdot\frac{\Delta\varphi}{2}\right)\cos^2\left(4\cdot\frac{\Delta\varphi}{2}\right) \tag{6.13}$$

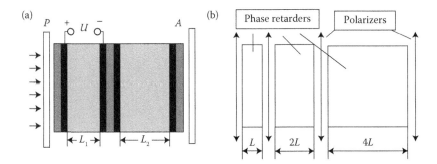

FIGURE 6.9   (a) Single-stage LCTF system and (b) three-stage Lyot interferential filters.

Tuning the wavelength of peak transmission in the Lyot PIF requires changing the path-length difference, or retardance, of each filter stage.

Application of an electric field in an LC device produces an analog variable retardation. It is therefore quite obvious how a passive PIF can be retrofitted to be active with nematic LC.

The wavelength corresponding to the maxim value of the transmission factor and the band pass of the hybrid tunable filter can be calculated with the relations (Dumitraşcu et al. 2008)

$$\lambda_{k,\max} = \frac{L_1 \Delta n_1 \pm L_2 \Delta n_2}{k} \quad \Delta\lambda = \frac{\lambda^2}{2(L_1 \Delta n_1 \pm L_2 \Delta n_2)} \tag{6.14}$$

Relations (6.5), (6.13), and (6.14) are the equations of the model used for simulating the Lyot-tunable LC–PIF transmittances shown in Figure 6.10. The simulation window displays two graphics (similar to those displayed in Figure 6.7—for comparison reasons) showing the transmittances versus the wavelengths corresponding to the two cases (Case I and Case II explained in Figure 6.6a and b—parallel optical axes and, respectively in Figure 6.6a and b—perpendicularly optical axes).

From Figure 6.10, one can see the same very clear difference in the half-pass band values which correspond to the two different orientations of the optical axis (in Case I, the half-pass band, around a wavelength of about 540 nm, is smaller than in Case II; this remark suggests [as in the case of Wood polarization interference filters] that the orientation corresponding to parallel optical axes of the two LC layer components of the basic stage is more efficient in filtering the optical radiations around the wavelength of 540 nm).

From Figure 6.10, one can also see the increasing filtering effect produced by the increased number of stages. From this point of view, the better results, regarding the half-pass band values, correspond to the five stages of Lyot PIF filter (TL-I_16 and TL-II_16).

6.5.5   Results and Discussions

The simulations of tunable polarization interferential filters made of LCs presented here allow a continuous modification of the wavelengths corresponding to the

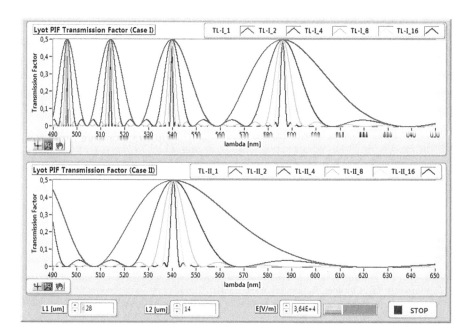

FIGURE 6.10   LabView application frame that simulates a single-stage and some multistage Lyot PIF transmission factor.

maximum value in the transmission band, through the modification of the electrostatic field intensity and the LC layer thicknesses, when the three scrollbars are used.

When the thicknesses L_1 and L_2 increase, the number of transmission bands obtained in the visible range increases while the band pass decreases.

The simulations prove the possibility to use the MBBA LC for making optoelectronic devices, such as a Lyot or a Wood multistage-polarizing interference filter. Similar simulations for LCTF, made of multistage lyotropic LCs, have been published in the specialized literature (Dumitraşcu et al. 2008).

The simulations prove that tunable filters with continuous control are generally fit for applications requiring a low-resolution level. In order to obtain higher resolutions (narrower transmission bands), it is required to increase the number of elements. The effect of the increase in the number of elements, from 3 to 5, can be seen in Figures 6.8 and 6.10.

The conditions in which the filters function, such as the value of the intensity of the electrostatic field, the thickness, or the optical features of the LC layers, may be easily set by simulation. The simulations may be used for the study and design of electrically controlled tunable interferential polarization filters. They allow to avoid conducting a large number of experiments needed in order to set the desired parameters of a certain filter.

REFERENCES

Born, M. and Wolf, E. 1999. *Principles of Optics*. Cambridge University Press, Cambridge, United Kingdom.

Bruhat, G. 1965. *Cours de Physique Generale. Optique*. Masson, Paris, France.

Demus, D., Goodgy, J.W., Gray, G.W., Spiees, H.W., and Vill, V. (Eds.). 1998. *Handbook of Liquid Crystals—Fundamentals*. Wiley-VCH, Weinheim.

Diaconu, I., Melniciuc, P.N., Dorohoi, D.O., and Aflori, M. 2007. Birefringence dispersion of *N*-(4-methoxybenzylidene)-4 butylaniline (MBBA) determined from channeled spectra. *Spectrochim. Acta Part A* 68:536–541.

Dorohoi, D., Postolache, M., and Postolache, M. 2001. Birefringence dispersion of poly (phenyl methacrylic) ester of cetyloxybenzoic acid in tetrachloromethane, determined from channeled spectra. *J. Macromol. Sci. Phys.* B40(2):239.

Dorohoi, D.O., Dumitraşcu, I., and Dumitraşcu, L. 2006. Optical transmission simulation of two anisotropic layers placed between two polarizers. *The 37th International Scientific Symposium of the Military Equipment and Technologies Research Agency (METRA)*, Bucharest.

Dumitraşcu, I. and Dorohoi, D.O. 2015. *Optical Anisotropy. Applications*. Tehnopress, Iaşi.

Dumitraşcu, I., Dumitraşcu, L., and Dorohoi, D.O. 2006. The influence of the external electric field on the birefringence of nematic liquid crystal layers. *J. Optoelectron. Adv. Mater.* 8(3):1028.

Dumitraşcu, I., Dumitraşcu, L., and Dorohoi, D.O. 2007. Polarimetry based on MBBA liquid crystalline layers as variable retarders. *AIP Conference Proceedings*, Istanbul, Turkey, 899:387.

Dumitraşcu, L., Dumitraşcu, I., and Dorohoi, D.O. 2006a. Computational modeling of the main refractive indices dispersion of the nematic liquid crystals in the visible range. *The 37th International Scientific Symposium of the Military Equipment and Technologies Research Agency (METRA)*, Bucharest, May 25th–26th.

Dumitraşcu, L., Dumitraşcu, I., and Dorohoi, D.O. 2006b. Thickness computation of the anisotropic layers used in the analysis of the light polarization state. *Materiale Plastice* 43(2):127–131.

Dumitraşcu, L., Dumitraşcu, I., and Dorohoi, D.O. 2007a. Hybrid tunable filters from MBBA and Island spath. A computer simulation. *Proceedings of BPU6*, August 22–26, Istanbul, Turkey.

Dumitraşcu, L., Dumitraşcu, I., and Dorohoi, D.O. 2007b. External electrostatic field influence on the order parameter of nematic liquid crystalline thin layers. *AIP Proc.* 899:389–390.

Dumitraşcu, L., Dumitraşcu, I., and Dorohoi, D.O. 2008. Tunable filters from lyotropic liquid crystals. *U.P.B. Sci. Bull., Ser. A (Bucharest)* 70(4):57–66.

Holmes, D.A. 1964. Exact theory of retardation plates. *J. Opt. Soc. Am.* 54:1115.

Lyot, B. 1933. Optical apparatus with wide field using interference of polarized light. *C.R. Acad. Sci. (Paris)* 197:1593.

Lyot, B. 1944. Filter monochromatique polarisant et ses applications en physique solaire [Monochromatic polarizing filters and their applications in solar physics]. *Ann. Astrophys.* 7:32.

Muscutariu, I. 1981. *Cristale lichide şi aplicaţii*. Tehnică, Bucureşti.

Ohman, Y. 1938. A new monochromator. *Nature* 41: 157, 291.

Ohman, Y. 1958. On some new birefringent filter for solar research. *Ark. Astron.* 2:165.

Picos, S., Amarandei, G., Diaconu, I., and Dorohoi, D. 2005. The birefringence of thin films of some nematic liquid crystals. *J. Optoelectron. Adv. Mater.* 7(2):787–793.

Pop, V. 1988. *Bazele Opticii*. Univ. Al.I.Cuza, Iaşi.

Savelyev, I.V. 1980. *Physics. A General Course Vol. II. Electricity and Magnetism, Waves Optics*. Mir Publishers, Moscow, English Translation.

Scharf, T. 2006. *Polarized Light in Liquid Crystals and Polymers*. John Wiley and Sons. Hoboken, New Jersey, USA.

Solc, I. 1965. Birefringent chain filters. *J. Opt. Soc. Am.* 55:621.

Xia, X., Stockley, J.E., Ewing, T.K., and Serati, S.A. 2002. Advances in polarization based liquid crystal optical filters. *SPIE Proceed.* 4658:51.

7 Novel Aspects on Optical Radiations Involvement in Analysis of Biaxial Crystals Optical Properties

Leonas Dumitrascu, Irina Dumitrascu, and Dana Ortansa Dorohoi

CONTENTS

7.1 INTRODUCTION

Progress in the field of optics in the last 50–60 years is due to remarkable discoveries and inventions among which one may mention

- The discovery of the laser effect and practical realization of a wide range of devices including semiconductor lasers
- The invention of optical fibers and their use in high-speed communications

- The discovery and synthesis of new materials with outstanding optical properties such as, for example, liquid crystals or various polymers used in the production of new optical components
- The realization of different types of microstructures and nanostructures, etc.

The application in practice of the results of these discoveries led to the increasing complexity of optical elements that form a part of the various types of apparatus or installations, and designing these elements became a more and more complex task.

Considering the fact that the operation of a large number of optical elements is based on the anisotropic properties of the material they are made, whatever the novelty of the recipe used in its production, it follows that the research of the optical radiation propagation through these types of media that possess such properties remains a topical issue.

Anisotropic plates with plane-parallel faces, such as, for example, phase plates (optical retarders) or special anisotropic plates (zero-order wave, half-wave, and quarter-wave plates), are optical elements that are found virtually in all devices that work with polarized optical radiation. Traditionally, these optical components are obtained by processing the anisotropic crystals (rough cutting followed by rectification, grinding, and polishing their surfaces) so that finally their faces are parallel to one of the principal coordinate system planes (Holmes 1964, Dumitraşcu et al. 2006c).

The adoption of this solution is based on the idea that the neutral lines of the anisotropic plate (directions of vibration) will be parallel with the two perpendicular axes of the main coordinate system, placed in the plane normal to the direction of propagation, the faces of the plate being normal to the third axis of the coordinate system.

This solution presents at least two major advantages:

- The use in optimal conditions (maximum) of the optical properties of birefringence-possessed material
- Simplicity and high precision of calculations design, because these calculations are based on the values of refractive indices of environmental lead

If one starts from analyzing the polyhedral shapes of the natural crystals (Figure 7.1), it is found that their orientation faces do not correspond generally to the orientation of crystallographic planes or to the main coordinate system of the anisotropic medium. For example, from Figure 7.1b and c, one can see that, in the case of natural crystals of quartz and calcite, the crystal faces are characterized by different sets of values $\{h, k, l\}$ of the Miller indices, and these sets do not correspond to the main crystallographic planes (Bloss, 1961).

In these circumstances, the obtaining of phase plates or other optical birefringent elements, whose faces contain one of the main coordinate system planes, may pose some disadvantages of which can be mentioned

- Increased consumption of material resulting from the need to eliminate a very large amount of raw material
- Increased volume of technological processing operations; these operations are both costly and difficult to perform in the case of high-hardness materials

(a)

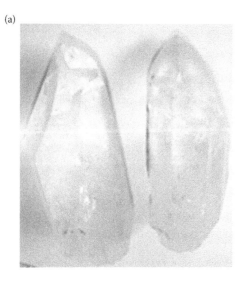

(b)   (c)

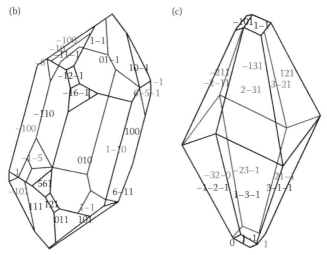

FIGURE 7.1   (a) Natural quartz crystals (Borşa - Baia Mare). (b) Miller indices of quartz crystal faces. (c) Miller indices calcite crystal faces.

Starting from these observations, the study must be focused on the following main objectives:

- Theoretically researching on the possibility of achieving birefringent optical components having one of the faces corresponding to one of the natural faces of the crystal block (the results can then be extended to the crystals obtained by artificial growth)
- Establishing the theoretical relations that allow for the design of such optical elements

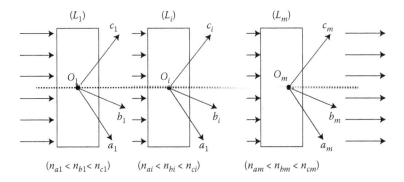

FIGURE 7.2 Schematic representation of electromagnetic harmonic plane wave propagation through a succession of anisotropic plates with plane parallel faces, cut obliquely with respect to the main coordinate system axes of the anisotropic materials.

- Testing the viability of the established theoretical relations by modeling and computational simulation carried out based on experimental data obtained from the study of optical properties of some natural anisotropic crystals (in this case, collected from Southern Carpathians, the mountain massive Lotru-Negoveanu)

Achieving these goals is closely linked to the problem of the plane harmonic electromagnetic waves propagation through a succession of anisotropic plates with plane-parallel faces, cut obliquely toward the axes of the main coordinate systems of the media, under normal incidence (Figure 7.2).

7.2   SEPARATION OF THE TRANSVERSE PROPAGATION MODES

To solve the above-mentioned problem, we start from the matrix representation of Maxwell's Equations (7.1)

$$
\begin{pmatrix}
i\omega\varepsilon_o\varepsilon_{rxx} & i\omega\varepsilon_o\varepsilon_{rxy} & i\omega\varepsilon_o\varepsilon_{rxz} & 0 & \dfrac{\partial}{\partial z} & -\dfrac{\partial}{\partial y} \\[2mm]
i\omega\varepsilon_o\varepsilon_{ryx} & i\omega\varepsilon_o\varepsilon_{ryy} & i\omega\varepsilon_o\varepsilon_{ryz} & -\dfrac{\partial}{\partial z} & 0 & \dfrac{\partial}{\partial x} \\[2mm]
i\omega\varepsilon_o\varepsilon_{rzx} & i\omega\varepsilon_o\varepsilon_{rzy} & i\omega\varepsilon_o\varepsilon_{rzz} & \dfrac{\partial}{\partial y} & -\dfrac{\partial}{\partial x} & 0 \\[2mm]
0 & -\dfrac{\partial}{\partial z} & \dfrac{\partial}{\partial y} & i\omega\mu_o\mu_{rxx} & i\omega\mu_o\mu_{rxy} & i\omega\mu_o\mu_{rxz} \\[2mm]
\dfrac{\partial}{\partial z} & 0 & -\dfrac{\partial}{\partial x} & i\omega\mu_o\mu_{ryx} & i\omega\mu_o\mu_{ryy} & i\omega\mu_o\mu_{ryx} \\[2mm]
-\dfrac{\partial}{\partial y} & \dfrac{\partial}{\partial x} & 0 & i\omega\mu_o\mu_{rzx} & i\omega\mu_o\mu_{rzy} & i\omega\mu_o\mu_{rzz}
\end{pmatrix}
\cdot
\begin{pmatrix}
E_x \\ E_y \\ E_z \\ H_x \\ H_y \\ H_z
\end{pmatrix}
= 0
$$

$$(7.1)$$

because this form of representation is naturally imposed by the tensor form of the constitutive relations for the anisotropic media and it allows a clear analysis of the situations in which these relations can be separated into the two simple transverse propagation modes of the electromagnetic wave (transverse electric [TE] and transverse magnetic [TM]); this separation is depending on (Kaminsky et al. 2004, Scharf, 2006):

- The type of the propagation medium, described analytically through the constitutive relations
- The propagation problem described in terms related to the number of dimensions of the propagation space

One can consider a plane harmonic wave of light that propagates in Oxy plane (Figure 7.3a). If the wave vector $\vec{k}$ is contained in the Oxy plane, it results that there are no variations in the intensity components of the position of the electric and magnetic fields in the direction of the axis Oz which means that all the derivatives of these components with respect to z are zero.

With regard to the propagation of transverse waves, they can be distinguished in the plane Oxy two particular modes (types) of propagation:

- The TE propagation mode, shown in Figure 7.3b, is characterized by the orientation of the electric component of the wave in the plane containing the direction of propagation.
- The TM propagation mode, shown in Figure 7.3c, is characterized by the orientation of the magnetic component of the wave in the plane containing the direction of propagation.

Because of the fixed orientation of the plane Oxy containing the direction of wave propagation, it results that the two propagation modes describe two linearly polarized waves having the vibration directions perpendicular to each other and being oriented at right angles to the direction of propagation.

In addition, from Figure 7.3b and c, one can notice that

- For the TE mode, only the components E_z, H_x, and H_y are different from zero.
- For the TM mode, only the components E_x, E_y, and H_z are different from zero.

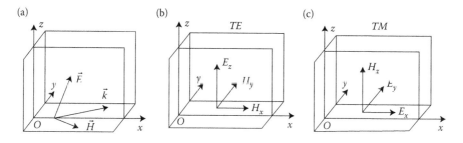

FIGURE 7.3 Illustrating: (a) Light wave propagation in Oxy plane, (b) TE propagation mode, and (c) TM propagation mode.

By analyzing the possibility of separating the TE and TM modes, we have shown that there are an infinite number of coordinate systems which correspond to the two-dimensional problem. The totality of the Cartesian coordinate systems, relative to which it is possible to separate the two TE and TM propagation modes, can be divided into three distinct classes with the following properties:

- Each class consists of an infinite number of representatives.
- All representatives of a class have in common one of the axes Oz, Oy, and Ox.
- The common axis for the representatives of each class coincides with one of the three main axes of the principal coordinate system of anisotropic environment (Oa, Ob, or Oc).

However, the approach based on the idea of separating the modes TE and TM is quite restrictive, if one takes into account the proposed objective, namely to describe in a uniform manner, propagation under normal incidence to a parallel beam through an indefinite sequence of plates with plane-parallel faces made from anisotropic media having the main axes in an oblique orientation with respect to the plates faces, because in the case of homogeneous and linear dielectric media, but anisotropic from the electrical point of view, the separation of the two simple propagation modes (TE and TM) is possible only for the two-dimensional problem.

7.3   DESCRIBING THE PROPAGATION MODES FROM THE COORDINATE SYSTEM OF CONSTANT PHASE SURFACE

An efficient approach to solve the three-dimensional problem is provided by the structure and characteristics of planar harmonic electromagnetic waves that propagate through anisotropic media, because during the propagation through an unspecified anisotropic medium, the electromagnetic waves of a certain wavelength consist of two modes of vibration, which, in the general case, they are not TE modes, but they are for the whole propagation distance through the medium-transverse vibration modes for the electric induction vector, with respect to the propagation direction of the surface of constant phase (transverse modes D).

7.3.1   CHARACTERISTICS OF PLANE ELECTROMAGNETIC WAVES IN IDEAL ANISOTROPIC MEDIA

Maxwell's equations of first order for the plane electromagnetic wave (Jenkins and White 1950) are

$$\begin{cases} \vec{k} \times \vec{E} = \omega \cdot \vec{B} \\ \vec{k} \times \vec{H} = -\omega \cdot \vec{D} \\ \vec{k} \cdot \vec{D} = 0 \\ \vec{k} \cdot \vec{B} = 0 \end{cases} \tag{7.2}$$

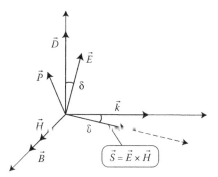

FIGURE 7.4   Structure of a plane electromagnetic wave in an anisotropic medium.

Equations 7.2 describe the structure of the plane electromagnetic wave that propagates through a homogeneous transparent and anisotropic dielectric medium; this structure is depicted in Figure 7.4.

From Equation 7.2 and Figure 7.4, it results in the following conclusions:

- The vectors $\vec{D}$, $\vec{E}$, $\vec{P}$, $\vec{k}$, and $\vec{S}$ are in the same plane, which is called the vibration plane of the wave.
- The vectors $\vec{B}$ and $\vec{H}$ (which are colinear in the case of isotropic media from the magnetic point of view) are perpendicular to the plane of vibration, and they define, together with the wave vector $\vec{k}$, the wave polarization plane.
- The plane defined by the vectors $\vec{D}$ and $\vec{B}$ ($\vec{H}$) represents in our case the surface wave plane, oriented perpendicular to the wave vector $\vec{k}$. (Vectors $\vec{D}$, $\vec{B}$ ($\vec{H}$), and $\vec{k}$, considered in that order, define a right-hand coordinate system.)
- The plane defined by the vectors $\vec{E}$ and $\vec{B}$ ($\vec{H}$) is the plane surface of energy transport, perpendicular to the Poynting vector $\vec{S}$. (One can notice that the triplet of vectors $\vec{E}$, $\vec{B}(\vec{H})$, and $\vec{S}$ also defines a right-hand coordinate system.)
- In general, the wave surface plane is different from the surface plane of energy transport; these planes form a dihedral angle whose measure is given by the angle δ between vectors $\vec{k}$ and $\vec{S}$. The fact that the angle δ is different from zero shows that the propagation direction of the wave surface is generally different from the direction of energy transport.

7.3.2   BASIC DIRECTIONS OF VIBRATION COORDINATE SYSTEM

For any direction of propagation through an anisotropic medium described by the wave vector $\vec{k}$, there are only two directions of vibration for electric induction $\vec{D}_1$ and $\vec{D}_2$ described by the unit vectors $\hat{D}_1$ and $\hat{D}_2$; these vibration directions can be regarded as those vibration directions capable of a given direction of propagation.

One can define the coordinate system of the basic vibration directions as the coordinate system $Oxyz$ having the following coordinate axes (Figure 7.5a):

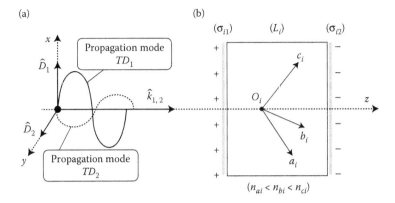

FIGURE 7.5 (a) The basic directions vibration coordinate system of the propagation medium. (b) The electric charge surface densities (σ_{i1}, σ_{i2}) on the two faces of the anisotropic plate L_i.

- The Oz axis is parallel to the propagation direction of the constant phase surface, given by the direction of the wave vector $\vec{k}$
- The Ox axis is parallel to the direction of vibration $\vec{D}_1$
- The Oy axis is parallel to the direction of vibration $\vec{D}_2$

In conclusion, the coordinate system of the basic vibration directions of the anisotropic propagation medium (Figure 7.5a) is a system of coordinates which can be considered for each given propagation direction; its axes have the directions described by the unit vectors $\hat{D}_1$ and $\hat{D}_2$ and by the unit vector of the propagation direction $(\hat{D}_1, \hat{D}_2, \hat{k})$.

Using the coordinate system of the basic vibration directions of the anisotropic propagation medium is advantageous because both propagation modes, TD_1 and TD_2, preserve their status (transverse modes D) even in the situation in which some anisotropic layers, shown in Figure 7.5b, would have electrical charges of densities σ_{i2} and σ_{i1} distributed on their surface (Figure 7.5b). This remarkable property follows from the fact that in the case of propagation modes TD, there is no longitudinal induction component; so, there are no jumps during the crossing of the separation surface between anisotropic media (in the case where such a surface area would have distributed electric charges).

To describe the propagation modes TD_1 and TD_2 from the basic vibration directions coordinate system of the propagation medium, the dielectric tensor components in relation to this coordinate system have to be known, because they represent the basis in the computation of slow and fast refraction indices of the medium for the considered direction.

7.4   DETERMINATION OF FRESNEL EQUATION ROOTS FOR BIAXIAL ANISOTROPIC MEDIA

To study the propagation of electromagnetic waves through the transparent anisotropic plates with plane-parallel faces, it is necessary (as it was shown in the previous section) to effectively know the general expression of the corresponding refractive indices of the two basic vibration directions as a function of the six parameters n_a, n_b, n_c, α, β, and γ is needed, and as well as those particular cases that may arise.

For this reason, the following is a detailed discussion of Fresnel's equation roots with their physical interpretation.

In the case of biaxial anisotropic media, the equation of the refractive index surface of Fresnel has the following form:

$$\frac{\alpha^2 n_a^2}{n_a^2 - n^2} + \frac{\beta^2 n_b^2}{n_b^2 - n^2} + \frac{\gamma^2 n_c^2}{n_c^2 - n^2} = 0 \tag{7.3}$$

in which the usual notations were made: n_a, n_b, and n_c represent the main refractive indices of the anisotropic media and, α, β, and γ are the director cosines of the wave propagation direction, described by the wave vector $\vec{k}$.

7.4.1 The General Case

This case represents a situation in which the director cosines α, β, and γ correspond to an arbitrary propagation direction, the only restriction being imposed by the condition $\alpha^2 + \beta^2 + \gamma^2 = 1$.

Making the notations

$$\begin{cases} A = \alpha^2 n_a^2 + \beta^2 n_b^2 + \gamma^2 n_c^2 \\ B = \alpha^2 n_a^2 (n_b^2 + n_c^2) + \beta^2 n_b^2 (n_c^2 + n_a^2) + \gamma^2 n_c^2 (n_a^2 + n_b^2) \\ C = n_a^2 n_b^2 n_c^2 \end{cases} \tag{7.4}$$

the real roots of Equations 7.3 are

$$\begin{cases} n_{1,2} = \pm \frac{1}{2}\sqrt{2}\sqrt{\frac{1}{A}\left(B - \sqrt{B^2 - 4AC}\right)} \\ n_{3,4} = \pm \frac{1}{2}\sqrt{2}\sqrt{\frac{1}{A}\left(B + \sqrt{B^2 - 4AC}\right)} \end{cases} \tag{7.5}$$

Analyzing relations (7.5), one can notice that the refractive indices group two by two after their absolute value.

This remark can be interpreted as follows:

• The fact that there are two distinct absolute values of refractive indices ($|n_{1,2}|$ and $|n_{3,4}|$, respectively) shows that, for a given propagation direction, there are generally two independent linearly polarized waves, which propagate with different speeds and which maintain their polarization state unchanged throughout the propagation process. The wave of which propagation is characterized by the minimum value of ($|n_{1,2}|$ and $|n_{3,4}|$) is the fast wave, and the wave characterized by the maximum value of ($|n_{1,2}|$ and

$|n_{3,4}|$) is the slow wave. It can be shown by using Maxwell's equations, that the polarization planes (or the vibration directions) of these two waves are always perpendicular to each other.

- The fact that there are two indices of refraction with the same absolute value, but with opposite sign ($n_1 = -n_2$ and $n_3 = -n_4$ respectively), shows that in each plane of polarization mentioned above, it may be simultaneously two waves that propagate with the same speed, but in opposite directions (progressive wave and regressive wave).

7.4.2 PARTICULAR CASES

Particular cases of propagation through a biaxial anisotropic medium can be obtained by particularizing the director cosine values of the propagation direction and they can be divided into two groups:

1. The cases in which the wave propagation direction is parallel to one of the axes Oa, Ob, or Oc of the main coordinate system.
2. The cases in which the wave propagation direction is contained in one of the main planes Oab, Oac, or Obc; for the waves propagating in the plane Oac, it should be highlighted in the special situation of the propagation along the optical axis of the medium.

Each group of particular cases mentioned above corresponds to three cases summarized in Table 7.1.

The wave propagation in the plane Oac ($\beta^2 = 0 \Rightarrow \alpha^2 + \gamma^2 = 1$) is of particular interest. In this case, analyzing the relations presented on the last row from Table 7.1, one can notice that the refractive indices have constant values, which mean that

---

TABLE 7.1

Synthetic Presentation of the Refractive Indices Values for Different Wave Propagation Directions through a Biaxial Anisotropic Media

Case	Roots $n_{1,2}$ and $n_{3,4}$
$\alpha^2 = 1 \Rightarrow \beta^2 = \gamma^2 = 0$	$n_{1,2} = \pm n_b$; $n_{3,4} = \pm n_c$
$\beta^2 = 1 \Rightarrow \alpha^2 = \gamma^2 = 0$	$n_{1,2} = \pm n_a$; $n_{3,4} = \pm n_c$
$\gamma^2 = 1 \Rightarrow \alpha^2 = \beta^2 = 0$	$n_{1,2} = \pm n_a$; $n_{3,4} = \pm n_b$
$\gamma^2 = 0 \Rightarrow \alpha^2 = 1 - \beta^2$	$n_{1,2} = \pm n_c$; $n_{3,4} = \pm \dfrac{n_a n_b}{\sqrt{n_a^2 + \beta^2 (n_b^2 - n_a^2)}}$
$\alpha^2 = 0 \Rightarrow \beta^2 = 1 - \gamma^2$	$n_{1,2} = \pm n_a$; $n_{3,4} = \pm \dfrac{n_b n_c}{\sqrt{n_b^2 + \gamma^2 (n_c^2 - n_b^2)}}$
$\beta^2 = 0 \Rightarrow \alpha^2 = 1 - \gamma^2$	$n_{1,2} = \pm n_b$; $n_{3,4} = \pm \dfrac{n_a n_c}{\sqrt{n_a^2 + \gamma^2 (n_c^2 - n_a^2)}}$

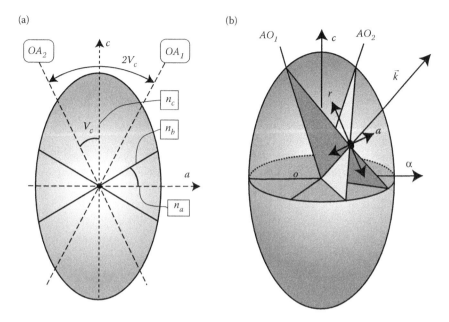

FIGURE 7.6 (a) The angle V_c used in classifying the biaxial anisotropic media, and (b) the refractive index ellipsoid.

one of the two independent linearly polarized waves, those from the main plane Oac, propagates always at the same speed, regardless of the direction of propagation relative to the axis Oc. Moreover, its vibration plane always contains the axis Ob, because $|n_{1,2}| = n_b$.

The second linearly polarized wave has a speed dependent on the angle formed with the axis Oc (Figure 7.6), because the absolute value of $|n_{3,4}|$ depends on the value of the director cosine γ. The vibration plane of this wave must be included in the main plane Oac (Bruhat 1965, Wright 1966).

The extreme values of the module $|n_{3,4}|$ are

- $|n_{3,4}|_{max} = n_c < n_b$ for $\gamma^2 = 0$ (propagation parallel to the axis Oa; this wave possesses the status of a slow wave, having the lowest propagation speed)
- $|n_{3,4}|_{min} = n_a > n_b$ for $\gamma^2 = 1$ (propagation parallel to the axis Oc; this wave possesses the status of a fast wave, having the highest propagation speed)

For the rest of the propagation directions from the plane Oac, the linearly polarized wave speed changes according to the angle formed by its propagation direction and the axis Oc; the propagation speed increases from the value that corresponds to the refractive index n_c to the value that corresponds to the refractive index n_a.

There are two special directions of propagation for which the propagation speed of the two waves propagating in the plane Oac becomes equal; these directions are called optical axes of the biaxial anisotropic medium, and the propagation speed along their directions has a value which corresponds to the n_b refractive index (Figure 7.6).

Solving the equation in relation to γ and denoting $V_c = \pm\arccos \gamma$, the following roots are obtained:

$$V_c = \pm\arccos\left(\frac{n_a}{n_b}\sqrt{\frac{n_c^2 - n_b^2}{n_c^2 - n_u^2}}\right) \tag{7.6}$$

The double value of the angle V_c, calculated with relation (7.6), represents and forms the basis for classification of biaxial anisotropic media.

Unlike the case of uniaxial crystals to which the optical sign is defined by means of birefringence, in the case of biaxial crystals, the optical sign is defined with the aid of the angle formed by the two optical axis, OA_1 and OA_2, with the bisector axis Oc (Figure 7.7a), as follows:

- If $2V_c < 90°$, the crystal is called biaxial positive (+).
- If $2V_c > 90°$, the crystal is called biaxial negative (−).
- If $2V_c = 90°$, the crystal is called biaxial neutral.

7.5 SOME EXPERIMENTAL DETERMINATION OF THE MAIN REFRACTIVE INDICES BY INTERFEROMETRIC MEANS

The knowledge of optical characteristics of substances is important for

- The design of optical systems composed of anisotropic layers
- Computer modeling and simulation of such systems

This paragraph is devoted to experimental results regarding the main refractive indices and the birefringence of some inorganic anisotropic crystals by the interferometric method, based on the use of a Rayleigh interferometer (Dumitraşcu et al. 2005).

Table 7.2 presents the main refractive indices, the birefringence Δn_{ac}, and the tangent of the angle formed by the optical axes of some biaxial crystals computed with Equation 7.6; these data can be found in references Androne et al. (2005, 2006).

Table 7.3 presents the values of the main refractive indices and the birefringence for more samples of mica (muscovite) from the Contu-Negovanu (Southern Carpathians) from the references (Southern Carpathians); see reference Androne et al. (2006).

From Tables 7.2 and 7.3, one can see that between the main refractive indices, n_a, n_b, and n_c satisfy the relation $n_a < n_b < n_c$ adopted by convention. In conclusion, for all the analyzed samples, the optical axes are lying in the Oac plane.

7.6 DETERMINING THE FAST AND SLOW REFRACTIVE INDICES AND THE BASIC VIBRATION DIRECTION OF ANISOTROPIC BIAXIAL CRYSTALS

Establishing the equations which permit the calculation of slow and fast refraction indices, and the angles that describe the orientation of the basic directions of vibration as a function of the main refractive indices of the medium and the angles defining

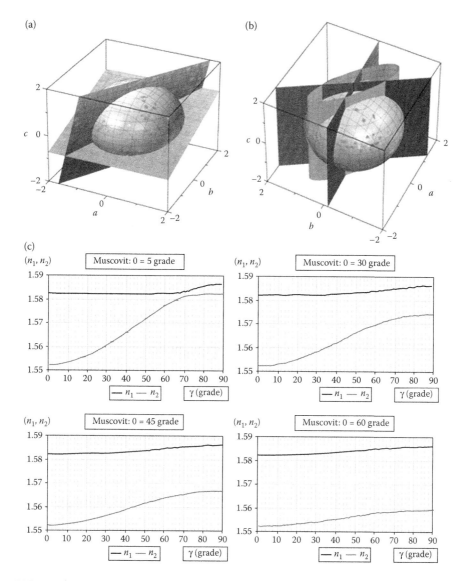

FIGURE 7.7 (a) Relative orientation of the basic vibration planes, (b) intersection of the basic vibration planes with the refractive index ellipsoid, and (c) slow and fast refractive indices of muscovite as a function of wave propagation direction.

the wave propagation direction for any direction of propagation (direct problem) is very useful for the following two reasons:

- They can be used to design the anisotropic plates with plane-parallel faces having a general orientation with respect to the main coordinate system, in particular for designing the special anisotropic plates.

TABLE 7.2

Main Refractive Indices, the Birefringence, and the Tangent of the Angle Formed by the Optical Axes of Some Biaxial Crystals

Sample	n_a	n_b	n_c	Δn_{ac}	$tg V$	V (rad.)
Feldspar	1.5360	1.5422	1.5482	0.0122	1.0225	0.7965
Albite	1.5275	1.5315	1.5375	0.0100	0.8205	0.6871
Mica	1.5601	1.5933	1.5940	0.0332	6.9989	1.4288
Aragonite	1.6859	1.6816	1.5301	−0.1558	6.3870	1.4154
Topaz	1.6221	1.6138	1.6112	−0.0109	0.5625	0.5124
Gypsum	1.5296	1.5226	1.5205	−0.0091	0.5501	0.5029

• They can represent a starting point in developing algorithms for computational modeling and simulation of various problems regarding the propagation of optical radiation through optical anisotropic media.

These equations are useful because they could be used in determining the orientation of the optical axes of the main coordinate system of a given anisotropic crystal with respect to the crystalline polyhedron faces, by determining the neutral lines of a crystalline plate with plane-parallel faces and the slow and fast refractive indices corresponding to the normal incidence.

If one only knows how the plate was cut (orientation of the plate faces relative to the axes of the main coordinate system) and the anisotropic material from which it is made, then, the slow and fast refractive indices and the orientation of the basic vibration directions of the medium must be determined by calculation.

TABLE 7.3

Main Refractive Indices and the Birefringence for Some Crystals of Mica (Muscovite) from the Contu-Negovanu (Southern Carpathians)

Sample	n_a	n_b	n_c	Δn_{ac}	Δn_{bc}	Δn_{ab}
Ms5	1.556	1.585	1.589	0.033	0.004	0.029
Ms7	1.555	1.585	1.600	0.045	0.015	0.030
Ms15	1.557	1.583	1.594	0.037	0.011	0.026
Ms17A	1.570	1.612	1.618	0.048	0.006	0.042
Ms22	1.562	1.584	1.591	0.029	0.007	0.024
Ms25	1.563	1.587	1.593	0.030	0.006	0.024
Ms71	1.568	1.610	1.617	0.049	0.007	0.042
Ms73	1.564	1.598	1.614	0.050	0.016	0.034
Ms75	1.567	1.602	1.611	0.044	0.009	0.035
Extreme values:	1.555–1.570	1.583–1.612	1.589–1.618	0.033–0.050		

7.6.1 EQUATIONS OF THE MODEL

In the case of biaxial anisotropic media, the solution of the propagation problem for the transverse modes TD_1 and TD_2 is carried by applying the Lagrange multipliers method. Using this method, calculation relations are established for

- The slow and fast refraction indices
- The direction cosines of the basic vibration directions as functions of the main refractive indices of the anisotropic media (n_a, n_b, and n_c) and direction cosines (α, β, and γ) corresponding to the propagation direction with respect to the main coordinate system axis

The method used to determine the values of slow and fast refractive indices along with the corresponding values of the cosine directors of the basic vibration directions, described by the orientation of the vectors $\vec{D}_1$ and $\vec{D}_2$, starts from the ellipsoid index and the idea that they must have extreme values (maximum and minimum) for any direction of propagation $\vec{k}$ (Born and Wolf 1999). By using the method of Lagrange multipliers, one can obtain the following result (Dumitraşcu and Dorohoi 2009):

$$
\begin{cases}
n_1 = \sqrt{n_{x_1}^2 + n_{y_1}^2 + n_{z_1}^2} \\
\alpha_1 = \dfrac{n_{x_1}}{\sqrt{n_{x_1}^2 + n_{y_1}^2 + n_{z_1}^2}} \\
\beta_1 = \dfrac{n_{y_1}}{\sqrt{n_{x_1}^2 + n_{y_1}^2 + n_{z_1}^2}} \\
\gamma_1 = \dfrac{n_{z_1}}{\sqrt{n_{x_1}^2 + n_{y_1}^2 + n_{z_1}^2}}
\end{cases}
\quad
\begin{cases}
n_2 = \sqrt{n_{x_2}^2 + n_{y_2}^2 + n_{z_2}^2} \\
\alpha_2 = \dfrac{n_{x_2}}{\sqrt{n_{x_2}^2 + n_{y_2}^2 + n_{z_2}^2}} \\
\beta_2 = \dfrac{n_{y_2}}{\sqrt{n_{x_2}^2 + n_{y_2}^2 + n_{z_2}^2}} \\
\gamma_2 = \dfrac{n_{z_2}}{\sqrt{n_{x_2}^2 + n_{y_2}^2 + n_{z_2}^2}}
\end{cases}
\tag{7.7}
$$

The simulation starts from the model given in Equations 7.7 and from experimental data regarding the main indices of refraction for a crystal of mica (muscovite) which is anisotropic biax. Refractive index values used for a small crystal were $n_a = 1.552$, $n_b = 1.582$, and $n_c = 1.586$.

In the simulation, the following two categories of elements are considered:

1. Graphical simulation of vibration basic directions relative to the direction of propagation and their relationship with refractive indices ellipsoid shape for some directions of propagation. Some simulation results of the vibration directions orientation are shown in Figure 7.7a and b, which illustrate the intersection between the basic vibration planes and the refractive index ellipsoid.

2. Numerical simulation results are shown in Figure 7.7c. These graphics represent the dependence of slow and fast refractive indices values as a function of propagation direction described by using the angles between the propagation direction and the axes of the main coordinate system.

7.7 OPTICAL MULTILAYER INTERFERENTIAL POLARIZATION FILTERS OF WOOD TYPE

The polarization interferential filters of Wood type are achieved from a transparent anisotropic layer of a constant thickness L placed between two polarizers P and A having their transmission directions parallel (Dorohoi et al. 2006) or crossed (Figure 7.8).

Their transmission factor varies with the optical radiations wavelength, having alternating maxims and minims, so that the transmitted radiation spectrum is alike to a spectrum (Pop 1988). The number of the transmitted maxims from a given spectral range depends on the birefringence and on the thickness of the anisotropic layer (Dumitraşcu et al. 2007).

When the angle between the basic directions of the anisotropic layer and the transmission directions of the polarization filters is $\theta = 45°$ (Figure 7.7) and neglecting the losses both by radiation absorption in the anisotropic layer and by reflection at the separation surfaces, the transmission factor of the device achieved from m identical elements is given by relation (7.8) (Dumitraşcu and Dorohoi 2009):

$$T_m(\lambda_o) = \frac{1}{2}\sin^{2m}\frac{\pi(n_a - n_c)L}{\lambda_o}, \quad (m = 2,3,4,\ldots) \tag{7.8}$$

The wavelengths corresponding to the maxims and minims of transmission and the number of the transmitted bands in the considered spectral range λ_1 and λ_2 ($\lambda_1 < \lambda_2$) are the same with those obtained with a filter achieved by one element (Dumitraşcu and Dorohoi 2009).

The transmission spectral bandwidth neighboring the maxim of k order is modified (one decreases). It is given by relation (7.9).

$$\Delta\lambda_{(m)} = \frac{4\pi(n_a - n_c)\arcsin(1/2^m)}{(2k+1)^2\pi^2 - 4(\arcsin(1/2^m))^2}L \tag{7.9}$$

Relations (7.8) and (7.9) are the equations used for simulation of the model Wood filter.

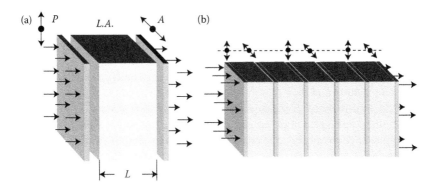

FIGURE 7.8   (a) Structure of a Wood filter element; (b) multilayer Wood filter achieved by $m = 5$ elements.

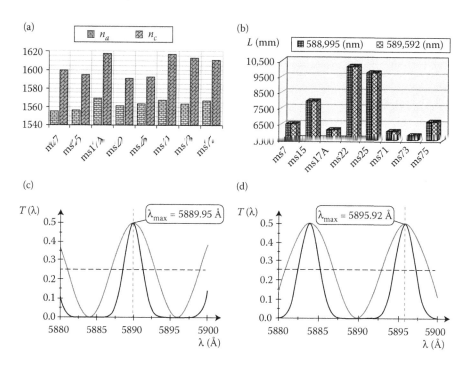

FIGURE 7.9 (a) Main refractive indices n_a and n_c of some micas (muscovite) samples, (b) mica layer thicknesses for the sodium doublet, and (c) and (d) transmission factor of a Wood multilayer ($m = 5$) filter separating the two neighboring lines of the sodium doublet.

The experimental data used in this simulation were determined by interferometric measurements in conformity with the ones described in Androne et al. (2006) and Dumitraşcu et al. (2006b). Some results referring to the main refractive indices for mica (muscovite) given in Androne et al. (2005) and Dumitraşcu et al. (2006b) are presented in Figure 7.9a; these values correspond to the wavelength of the sodium yellow doublet (5889.95 Å and 5895.92 Å).

The filters were projected so that they transmit only one from the sodium doublet lines and to block the second line (Dumitraşcu and Dorohoi 2009). From Figure 7.9b, it results that the mica layer thicknesses necessary to eliminate one from the two neighboring lines are of the order of millimeters (the interval 5.5 and 10.5 mm).

The transmission factors of some Wood filters achieved by $m = 5$ layers of the type ms7 are drawn with thick lines in Figure 7.9c and d. To illustrate the diminishing of the transmitted band by a multilayer ($m = 5$) Wood filter compared with a monolayer filter ($m = 1$), the transmission factors of the monolayer filters were drawn with a thin line. The bigger the number of the used anisotropic layer, the thinner is the passing band of a Wood multilayer polarization interferential filter. From the considered simulation (the dotted line in Figure 7.9c and d), it results that for $m = 5$, the transmission band is diminished at half compared to the case when $m = 1$.

One important disadvantage of such kind of filters is the fact that they must be separately projected and achieved for each pair of spectral lines which must be separated. It can be eliminated by using interferential tunable polarization filters whose transmission factor can be controlled with an external electric field applied to an anisotropic layer.

REFERENCES

Androne, D.A.M., Dumitraşcu, L., and Dorohoi, D.O. 2005. Inorganic crystals studied by optical means: I. Optical study of pegmatitic micas (muscovite). Annals of "Dunărea de Jos" University, Galaţi, Fasc. II, an XXIII (XXVIII), pp. 137–142.

Androne, D.A.M., Dumitraşcu, L., Horga, I., and Dorohoi, D.O. 2006. Main refractive indices and birefringence of feldspars from Romanian pegmatites. *Bull. Politechnical Inst. Iasi* 55(3&4):137–141.

Bloss, F.D. 1961. *An Introduction to the Methods of Optical Crystallography*. Holt, Rinehart and Winston, New York.

Born, M. and Wolf, E. 1999. *Principles of Optics*. Cambridge University Press, Cambridge, United Kingdom.

Bruhat, G. 1965. *Cours de Physique Generale, Optique*. Masson, Paris, France.

Dorohoi, D.O., Dumitraşcu, I., and Dumitraşcu, L. 2006. Optical transmission simulation of two anisotropic layers placed between two polarizers. *The 37th International Scientific Symposium of the Military Equipment and Technologies Research Agency (METRA)*, Bucharest.

Dumitraşcu, L. and Dorohoi, D.O. 2009. *Elemente de optica mediilor anizotrope. Aplicatii.* Tehnopress, Iasi.

Dumitraşcu, L., Dumitraşcu, I., and Dorohoi, D.O. 2005. Birefringence of the inorganic crystals, spectroscopy of molecules and crystals, *XVII International School-Seminar*, Sept 20–26, Beregove, Crimea, Ukraine.

Dumitraşcu, I., Dumitraşcu, L., and Dorohoi, D.O. 2006a. Optical study of some inorganic biax crystals. *Proceedings of BPU6*, August 22–26, Istanbul, Turkey.

Dumitraşcu, L., Dumitraşcu, I., and Dorohoi, D.O. 2006b. Determination of the basic vibration direction and the optical sign computation for biaxial crystals. *Proceedings of the National Session "Strategii XXI"*, Universităţii Naţionale de Apărare "Carol I", Bucureşti, pp. 183–191.

Dumitraşcu, L., Dumitraşcu, I., and Dorohoi, D.O. 2006c. Thickness computation of the anisotropic layers used in the analysis of the light polarization state. *Materiale Plastice* 43(2):127–131.

Dumitraşcu, L., Dumitraşcu, I., and Dorohoi, D.O. 2007. Hybrid tunable filters from MBBA and Island spath. A computer simulation. *Proceedings of BPU6*, August 22–26, Istanbul, Turkey.

Holmes, D.A. 1964. Exact theory of retardation plates. *J. Opt. Soc. Am.* 54:1115.

Jenkins, F.A. and White, H.E. 1950. *Fundamentals of Optics*, 2nd edition. McGraw-Hill, New York.

Kaminsky, W., Claborn, K., and Kahr, B. 2004. Polarimetric imaging of crystals. *Chem. Soc. Rev.* 33:514–525.

Pop, V. 1988. *Bazele Opticii*. Editura Universităţii, Iasi, Romania.

Scharf, T. 2006. *Polarized Light in Liquid Crystals and Polymers*. John Wiley and Sons, Hoboken, New Jersey, USA.

Wright, H.G. 1966. Determination of indicatrix orientation and 2V with the spindle stage: A caution and a test. *Am. Mineral.* 51:919–924.

Section II

Interactions of Electromagnetic
Radiations with
Macromolecular Materials

8 Designing Antimicrobial Properties on Urinary Catheter Surfaces by Interaction with Plasma Radiation and Particles

Magdalena Aflori

CONTENTS

8.1 IMPACT OF THE STUDY

The development of current healthcare practices is given by the emergence of new surgical procedures, materials, and medical products. Polymeric materials are present in many of these innovations and certain clinical and cost requirements must be met. The high performance of biomaterials is obtained by controlling surface properties (Aflori et al. 2013). The radio frequency (RF) plasmas are generally used for the surface modification of polymers because they produce the sterilization of the products, involve low consumption of chemicals, and it is a solvent-free and dry process. By using precisely controllable processes from RF plasma, the surface can be treated homogeneously and the surface chemistry can be tailored for the required end use (Drobota et al. 2010, Aflori 2014). In order to achieve antimicrobial surfaces properties, the research efforts have been turned toward engineering polymer films loaded with silver and silver nanoparticles (Johnson et al. 1990, Sagripanti 1992). Multilayer films containing silver nanoparticles were generated by layer-by-layer assembly, comprising catechol used to ligate and reduce silver ions to metallic silver when immersed in an aqueous silver salt solution (Hwang et al. 2011). Different commercial products containing silver have materialized: wound dressings, bandages for burns and chronic wounds, silver and silver nanoparticles-coated catheters, and other medical devices. A number of factors, such as concentration, size, shape, and aggregation, come into play in determining the efficiency of silver-based coatings (Liedberg and Lundeberg 1989, Ahearn et al. 2000). One of the most common types of healthcare-associated infection from hospitals is bacteriuria and is owing to the presence of a urinary catheter in the human body, the daily incidence of those urinary tract infections being 3%–10%. The mucosal irritation and catheter-related urinary tract infection (CUTI) are due to the insertion of the urethral catheters which inoculate organisms into the bladder and promote colonization by providing a surface for bacterial adhesion (Vergidis and Patel 2012, Padawer et al. 2015). A CUTI is defined as the new appearance of bacteriuria or funguria with a count of greater than 103 colony-forming units per milliliter (Patel and Arya 2000). Bacteriuria is developed in 10%–30% of patients who are undergoing short-term catheterization (i.e., 2–4 days) while in the case of long-term catheterization, the percents are 90%–100% (Gould et al. 2010, Hooton et al. 2010). About 80% of nosocomial CUTIs are related to urethral catheterization but only 5%–10% are related to genitourinary manipulation. The presence of potentially pathogenic bacteria and an indwelling catheter produce a nosocomial CUTI. The bacteria may gain entry into the bladder during manipulation of the catheter or drainage system, during insertion of the catheter, around the catheter, and after removal (Stark and Maki 1998, Darouiche et al. 1999). The most commonly responsible for CUTI development are enteric pathogens (e.g., *Escherichia coli*), but *Pseudomonas species, Enterococcus species, Staphylococcus aureus,* coagulase negative staphylococci, *Enterobacter* species, and yeast also are known to cause infection (Riley et al. 1995). Nowadays, there are many studies comparing standard noncoated catheters with silver-coated catheters, and all studies have included patients requiring a catheter *in situ* for more than 2 days and less than 10 days (Munasinghe et al. 2001, Schumm and Lam 2008). A complete analysis consisting of eight clinical trials was performed (Saint and Lipsky 1999) in order to establish the effect of silver coatings. When all studies were considered together (four of them using silver-oxide-coated catheters and the other four using silver-alloy-coated hydrogel catheters), a statistically significant benefit was seen in all patients with silver-coated catheters. Two randomized prospective studies were performed in order to evaluate the incidence of bacteriuria during short- and intermediate-term catheterization with the silver-alloy-coated hydrogel catheter compared to the standard latex catheter following urological surgery (Verleyen et al. 1999). The authors found that the silver alloy catheter significantly delayed the onset of bacteriuria in patients catheterized for a median duration of 5 days (range 2–14 days).

By day 7, 30% of patients with latex catheters and 10% with silver-coated catheters had significant bacteriuria. They found no difference in the incidence of CUTIs in patients following intermediate-term catheterization for 14 days (50% vs. 50.6%). A study on the catheterized patients from 12 different centers was made comparing infection rates between those receiving standard noncoated catheters and silver-coated hydrogel catheters. The use of the silver-alloy-coated catheter significantly reduced the risk of CUTIs from 32 to 20.4 CUTIs per 1000 catheterized patients: a 33% reduction in the number of patients with secondary bacteriuria and a 42% lower incidence of multidrug-resistant organisms (Thomas et al. 2002). A change in practice to silver-alloy-coated hydrogel catheters throughout the hospital was implemented, resulting in a 45% reduction in CUTIs (Lai and Fontecchio 2002). The use of standard catheters resulted in approximately 516 CUTIs per year compared to 300 CUTIs when using silver-alloy-coated catheters, thereby reducing the incidence by 216 per year.

Here, antimicrobial properties on urinary catheter surfaces commonly used in hospitals were obtained. To achieve this objective, a two-step method combined plasma and $AgNO_3$ wet treatments were performed. In order to obtain information about the chemical and morphological modifications of polymers, Fourier transform infrared–attenuated total reflectance (FTIR–ATR), scanning electron microscopy (SEM) with energy-dispersive x-ray analysis (EDXA), and x-ray diffraction (XRD) were used.

The sensitivity of *Pseudomonas* (one of the most common organisms associated with biofilm growth on catheters) to silver ions attached to the catheter surface was studied by performing culture media and inoculation of isolated bacterial strain at the surface of the silver-treated catheters.

8.2 PLASMA INTERACTION WITH SUBSTANCE

8.2.1 General Processes at the Interface between the Plasma and the Material

Plasma, the "fourth state of matter," is defined as a partially or wholly ionized gas with a roughly equal number of positively and negatively charged particles. There are two types of plasma—high temperature and low temperature. For high-temperature plasma, an example of naturally occurring plasma is lightning, while artificially generated arc examples are the corona discharge and the plasma torch, used to vaporize and redeposit metals. Low-temperatures plasmas are ionized gases generated within a vacuum chamber where atmospheric gases have been previously evacuated below 0.1 torr and the working gas is introduced at pressures between 0.1 and 2 torr. At low pressure, a relatively long free path of accelerated electrons and ions is obtained; therefore, these types of plasma are used in organic cleaning and surface modification. The ions and neutral particles are at or near ambient temperatures and the long free path of electrons has relatively few collisions with molecules at this pressure, so that the reaction remains at low temperature.

Plasma processing can produce unique effects of commercial value that can be obtained in no other way (Tatoulian et al. 1995), used in a large area of industrial applications and is preferable to conventional processes that accomplish similar results because

- Contributes to the reduction of carbon dioxide emissions and global warming and in energy consumption.
- Concurs to the increases in the efficiency of energy use and to the reduction of pollution and environmental contaminants can result in significantly reduced inputs.
- Can reduce unwanted by-products and occupational hazards at the point of manufacture.
- Can minimize toxic wastes.
- The active species are generated by the interaction of the working gas(es) with the plasma and are rarely available in purely chemical reactors in the concentrations and active states of excitation found in plasma reactors.

The working gas is energized by one of the following types of energy: RF, microwaves, and alternating or direct current. The energetic species in gas plasma include ions, electrons, radicals, metastables, and photons in the short-wave ultraviolet (UV) range. There are the processing parameters (such as gas types, power, time, and operating pressure) which can be varied by the user and the system parameters (such as electrode location, reactor design, gas inlets, and vacuum) which are set by the design of the plasma equipment. The wide range of the above-mentioned parameters offers greater control over the plasma processes. The most high-energy radiation processes are obtained due to the broad range of parameters. Surfaces in contact with the gas plasma are bombarded by the plasma-energetic species and their energy is transferred from the plasma to the solid and dissipated within the solid by a variety of chemical and physical processes. A unique type of surface modification that reacts with surfaces in depths from several hundred angstroms to 10 μm without changing the bulk properties of the material is obtained. The advantages of plasma treatment include low environmental impact, no line-of-sight problem as compared to laser and UV radiation, a pinhole-free coating, and continuous online treatment of fibers, tubing, membranes, fabrics, and films (Hsieh et al. 2009, Pandiyaraj et al. 2008).

8.2.2 Interaction with a Polymeric Surface

When the polymeric substrate is exposed to the action of a gas plasma, the plasma species interact with the substrate chemically and physically (Figure 8.1).

The collisions of the energetic species from plasma at the polymeric surface lead to the formation of the free radical by chain scission of molecules (Kauling et al. 2009), and then new functional groups being generated. The nature of the interactions between the excited species and the solid surface will determine the type and degree of the chemical and physical modifications that will take place. The processing conditions, such as power, pressure, gas, etc., and the nature of the substrate will determine whether the surface modification is one of film deposition, substitution, or ablation. In plasma treatment, gases that do not fragment into polymerizable

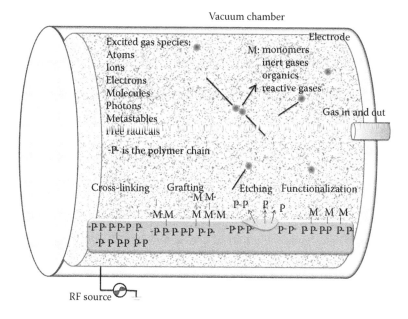

FIGURE 8.1 Plasma interactions with polymer substrates.

intermediates upon excitation are used. These include air, nitrogen, argon, oxygen, nitrous oxide, helium, tetrafluoromethane, water vapor, carbon dioxide, methane, and ammonia. Exposure to such plasmas can lead to the introduction of chemical functionalities, with the nature of the functionalities being highly dependent on the chemical composition of the biomaterial and the process gas. For instance, plasma oxidation, nitration, hydrolyzation, or amination will increase the surface energy and hydrophilicity of the biomaterial, therefore changing the way in which the biomaterial interacts with its immediate physiological environment (Subbiahdoss et al. 2009, Kwok et al. 2010).

The inert gases used in plasma treatment are hydrogen, helium, neon, and argon. Physical modification is the main effect on the surface by inert gas plasma treatment. Regardless, oxygen-related functional groups (−COH, −C=O, and −COOH) were introduced by inert gas plasma treatment. In oxygen plasma, the various active particles of oxygen molecules can be obtained by dissociation and combination reactions in oxygen plasma, while the inert gas plasma has ions and electrons. However, the radicals generated by plasma can interact with oxygen and H_2O in air after plasma exposure, and then, hydrophilic functional groups can be introduced on substrate surface (Biederman and Slavínská 2000, Desmet et al. 2009). In general, oxygen plasma treatment introduces various oxygen functional groups such as −CO−, −C=O, −COOH, −C−O−O−, and O−COO− on the polymer surface, resulting in wettability enhancement. Also, carbon dioxide plasma can lead to surface oxidation by forming oxygen-containing hydrophilic groups. Nitrogen plasma introduces nitrogen-containing functional groups such as amine, imine, and amide groups on polymer surface (Johnston and Ratner 1996, Denes and Manolache 2004). It was found that polystyrene surface treated by nitrogen plasma was highly reactive with oxygen after exposure to air. Ammonia plasma is also used to generate nitrogen-containing functional groups, improving wettability, printability, and adhesion (Yang et al. 2009, Inbakumar et al. 2010).

After plasma treatment on the substrate, unstable free radicals would recombine rapidly with other active species while stable free radicals remain as living radicals (Leroux et al. 2006, Gristina et al. 2009). Moreover, the chain cross-linking is induced by recombination between radicals. The activity of the plasma creates a higher cross-linking density within the material to depths of a few thousand angstroms; so, *cross-linking* is often used to increase the surface hardness and chemical resistance which can enhance the performance of polymers in many applications (Terlingen 1993, Anders 2005). By using a plasma immobilization process, pre-coated molecules can be directly cross-linked onto polymer surfaces. The molecules immobilized by this process can be organic compounds, surfactants, polymers, or proteins, and do not require unsaturated double bonds in the molecules. There are many other types of coatings that can be similarly enhanced (Chu 2007, Jacobs et al. 2010).

The radical formation processes are summarized as follows:

Process 1—Radical formation by abstraction of atoms

1. Elimination of atoms on the backbone of molecules of polymer surface by radicals generated from gas molecules in the plasma
2. Generation of carbon radicals on polymer surface
3. Chemical combination of carbon radicals with other radicals in the plasma
4. Functionalization—Formation of functional groups on the polymer surface

Process 2—Radical formation by chain scission

1. Electron and ion bombardments on polymer chains
2. Chain scission of polymer chain and formation of radicals at the end of polymer chains
3. Degradation products generated leading to the reduction of molecular weight on the polymer surface
4. Chemical combination of carbon radicals with other radicals in the plasma
5. Functionalization—Formation of functional groups on the polymer surface

The surface functionalities that arise as a result of plasma treatment can serve as a platform for further surface modification processes, such as the grafting of biomolecules and other functional structures. Further surface modification can be performed in order to tailor the properties of the biomaterial to a specific application (Sohbatzadeh et al. 2010, Sureshkumar et al. 2010).

Plasma surface treatment enhances the surface energy, wettability, wickability, and bonding of fabrics, films, and solids, and the sterilization, cleaning, and decontamination of surfaces. The used plasmas must have a sufficiently high electron number density to provide useful fluxes of active species, but not as high or energetic as to damage the material treated in order to achieve industrially important effects.

Thus, glow discharge plasmas operated at 1 atm or under vacuum possess the appropriate density and active-species flux for nearly all plasma surface treatment applications (Roth et al. 2010).

Although many biomaterials have physical properties that meet and even exceed those of natural body tissue, they can often cause adverse physiological reactions such as infection, inflammation, and thrombosis formation. Through surface modification, biocompatibility as well as biofunctionality can be achieved without changing the bulk properties of the material, plasma surface modification provides device manufacturers with a flexible, safe, and environmentally friendly process that is extremely effective. Designing a coating that will be effective in controlling bacterial adhesion and proliferation requires an in-depth understanding of the forces that govern these processes, the attachment and colony formation dynamics, and the consequences for both the colonizer and the abiotic target of adhesion (Raynor et al. 2009). Furthermore, the development of biomaterial-associated infections can arise in several ways, the most common being the introduction of etiological agents from direct contamination of the implant during surgery or postoperative care. In addition, microorganisms that originate from an infection site elsewhere in the body can spread through the blood, causing late hematogenous infection of the implant, particularly when medical devices are directly exposed to the blood (Huang et al. 2003).

8.3 EXPERIMENTAL PROCEDURE

8.3.1 TWO-STEP TREATMENT

This chapter describes a two-step method for obtaining silver-containing catheter surfaces for antimicrobial purposes. The catheters were treated first by argon and argon–oxygen plasma in an Emitech K1050X device. The plasma device has 110-mm-diameter × 155-mm-deep quartz chamber horizontally mounted with a convenient slide-out specimen drawer and viewing window. Urinary catheters were positioned horizontally on a metal sample holder. In each run, prior to the treatment, the reactor is evacuated down to a base pressure of 0.3 torr. Automatic tuning of forward and reflected power is standard. Forward power and vacuum levels are shown by the digital display. The treatments were performed in argon and argon discharges at an input power of 80 W and a discharge time of 6 min.

After plasma treatments, the catheters underwent chemical wet treatments in a sodium hydroxide (NaOH) and silver nitrate ($AgNO_3$) solution for 1 h under UV light action (Gheorghiu et al. 2005, Aflori et al. 2008).

8.3.2 TYPES OF MEASUREMENTS

The exposed samples were taken out at different time intervals, and the measurement of the absorption spectra was carried out. The modification in the catheter surface structure was examined by FTIR–ATR spectroscopy in the range of 600–5000 cm^{-1}. Peak height measurements were performed with the spectral analysis software (Opus 5).

SEM was used to analyze the surface morphology and topography. Prior to SEM analysis, untreated and modified substrates were sputter coated with an ultrathin layer of gold (JEOL-JFC) for 30 s at 20 mA. Afterward, samples were analyzed by SEM (JEOL-JSM-6400F) at an accelerating voltage of 15 kV.

Following XRD method, the x-rays were generated by using a Cu Kα source from a D8 Bruker diffractometer with an emission current of 36 mA and a voltage of 30 kV. Scans were collected over the 2 h = 2–60° ranges using a step size of 0.01° and a count time of 0.5 s/step. The analysis of the diffractograms was performed with an EVA soft from DiffracPlus package and the ICDD-PDF2 database.

The sensitivity of *Pseudomonas*-isolated bacterial strain to the silver ions attached to the catheter surface was tested. The bacterium was stored at −80°C in 10% glycerol. Bacteria were refreshed on Sabouraud dextrose agar (SDA) (Biokar, France) and were grown at 36°C. Using these cultures, microbial suspensions were prepared in sterile saline to obtain turbidity optically comparable to that of the 0.5 McFarland standards yielding a suspension containing approximately 1×108 CFU mL^{-1} for all microorganisms. Volumes of 0.2 mL from each inoculum were spread onto Mueller–Hinton Agar prepared in Petri plates. The antimicrobial properties were done for 2 months after performing the treatment, in order to evidence the stability of the treatments.

8.4 RESULTS AND DISCUSSIONS

8.4.1 FTIR MEASUREMENTS

The changes in the chemical structure of the catheter films were evidenced through FTIR–ATR spectroscopy analysis, which can be considered a valuable method for the characterization of polymer surface (Figure 8.2).

The presence of latex from the catheter surface was by the band at 1089 cm^{-1} and by the unsaturated carbon–carbon double bonds associated from latex molecules: 1378 cm^{-1} for C—H bending of CH_3, 2960 cm^{-1} for C—H stretching of CH_3, 1445 cm^{-1} for C—H bending of CH_2, 835 cm^{-1} for C=C wagging, and 1650 cm^{-1} for

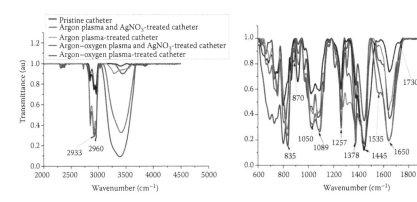

FIGURE 8.2 FTIR spectra for all studied samples.

C=C stretching. The side chain vibrations include aliphatic C—H stretching modes at 2933 cm^{-1}. The changes in FTIR spectra after plasma treatments are identified in Figure 8.2. The strong broad peak at 3400–3420 cm^{-1} is characteristic to OH stretching. This peak was shifted to higher wavenumbers due to the formation of peroxide groups around 1730 cm^{-1}.

The FTIR spectra of silver-containing catheter Ag show a strong absorption band at 1650 cm^{-1}. The band at 1445 cm^{-1} was assigned to methylene-scissoring vibrations. This peak also became sharp and increased in intensity, and shifted to lower wavenumbers when the sample was treated in plasma. The significant changes observed for peaks at 1378 and 1455 cm^{-1} are indicative of the role of aliphatic C—H groups in the reduction of Ag ions. The two peaks are closely merged in pure latex but are more individualized after interaction with silver. As for AgNO$_3$, the following absorption peaks were evidenced: 1535 cm^{-1} for N—O stretching, 1050 cm^{-1} for C—OH, C—O stretching, and for 870 cm^{-1} C—NO$_2$ stretching. The two peaks at 1050 cm^{-1}, 870 cm^{-1} are closely merged on latex, but they are more individualized after interaction with silver.

8.4.2 SEM Measurements

SEM image demonstrates a relatively flat surface of pristine catheter with few imperfections (Figure 8.3a).

The plasma treatments produced "hills and valleys surfaces" evidenced by SEM images (Figure 8.3b and c). Following silver reduction onto positron-emission tomography (PET) surface, the SEM images highlight the distribution of silver ions, evidencing a surface structure with randomly distributed grains plasma-treated catheter (Figure 8.3d); argon plasma followed by AgNO$_3$-treated catheter (Figure 8.3e); and argon–oxygen plasma followed by AgNO$_3$-treated catheter.

EDXA analysis (Table 8.1) was used to determine the element concentrations present on the surface. The carbon and oxygen were present in all measurements.

8.4.3 XRD Measurements

The XRD peaks of silver-treated urinary catheter (Figure 8.4) indicate the presence of silver (ICDD card 00-001-1164) with characteristic peaks at 29.45, 38.18, and 44.38; silver oxide (ICDD card 00-019-1155) with characteristic peaks at 33.66, 36.33, and 38.39; and a small amount of silver nitrite (ICDD card 00-001-0823) with characteristic peaks at 22.49, 29.17, and 46.03. The increasing of crystallinity degree (Table 8.2) demonstrates the restriction of catheter degradation to the amorphous regions and the chain scission of tie segments between crystallites in polymer

8.4.4 Antimicrobial Activity

Culture media and inoculation was performed in order to test the sensitivity of *Pseudomonas*-isolated bacterial strain, to silver ions attached to the catheter surface (Figure 8.5). The pristine catheter shows no antibacterial activity (Figure 8.5a),

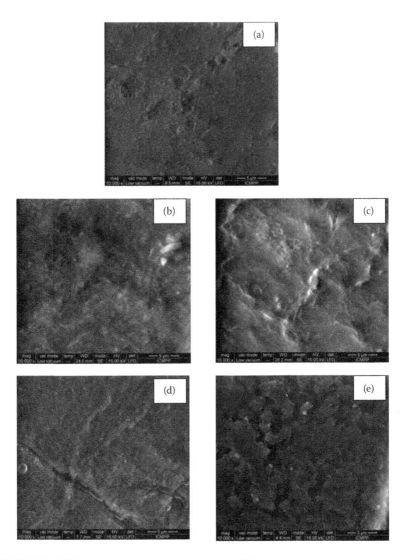

FIGURE 8.3   SEM images of (a) pristine catheter; (b) argon plasma treated catheter; (c) argon–oxygen.

while in case of the catheter treated first with argon plasma and followed by wet treatments, the shadow diameter is 7 mm (see arrows from Figure 8.5b), and in the case of the catheter treated first in argon–oxygen plasma and then in the solution of NaOH and AgNO$_3$ under UV action, the shadow diameter was 11 mm (see arrows from Figure 8.5c). The different dimensions for the inhibition zone demonstrate the influence of plasma treatment on the bactericidal activity of the urinary catheter. The solution containing silver has the same concentration in both cases, but in the case of the argon–oxygen plasma as a precursor for the wet treatment, the quantity of silver

TABLE 8.1
Atomic Concentration of C, O, N, and Ag (%) on the Surface of the Samples Calculated from the EDXA Measurements

Sample/EDXA	C	Si	O	Ag	N
Pristine catheter	81.03	4.82	14.15		
Argon plasma-treated catheter	84.12	2.80	13.08		
Argon–oxygen plasma-treated catheter	83.05	3.90	13.05		
Argon plasma and $AgNO_3$-treated catheter	82.04	1.80	11.06	4.08	1.02
Argon–oxygen plasma and $AgNO_3$-treated catheter	82.01	2.15	10.15	4.91	0.78

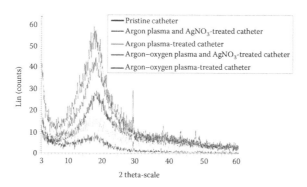

FIGURE 8.4   XRD data of all samples.

TABLE 8.2
Crystallinity Degree Calculated from XRD Diffractograms

Sample	Crystallinity Degree (%)
Pristine catheter	20
Argon plasma-treated catheter	27
Argon–oxygen plasma-treated catheter	42
Argon plasma followed by $AgNO_3$-treated catheter	31
Argon–oxygen plasma followed by $AgNO_3$-treated catheter	38

(see Table 8.1) is slightly higher than in the case of the argon plasma treatment as a precursor for the silver-containing solution treatment.

Silver is the broadest-spectrum antibiotic available, and it generally does not induce resistance (Grandin and Textor 2012). While contradictions exist in the literature (Vasilev et al. 2009), the general view is that metallic silver has little activity (despite reference to contact activation). The biological functions are exerted by the silver ions released from silver metal coatings: ions interact with bacterial cell walls,

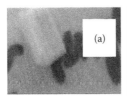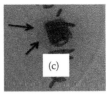

FIGURE 8.5   Culture media and inoculation to test the sensitivity of *Pseudomonas*-isolated bacterial strain for (a) pristine catheter; (b) argon plasma followed by $AgNO_3$-treated catheter; and (c) argon–oxygen plasma followed by $AgNO_3$-treated catheter.

complex with proteins and deoxyribonucleic acid (DNA), and affect their biological functions. Different scenarios have been proposed for the mode of action of silver against bacteria:

- Silver ions bind to the bacterial cell membrane and damage it by interfering with membrane receptors and bacterial electron transport (impeding the production of adenosine triphosphate, the cell's energy source).
- Silver ions bind to bacterial DNA and thus damage their replication, and cause intracellular formation of insoluble compounds with nucleotides and proteins; particularly, the amino acids histidine and cysteine will readily complexate with Ag^+ ions.
- Silver ions bind to sulfhydryl groups inactivating the respiratory chain enzymes, forming hydroxyl radical, and ensuing DNA damage (Gordon et al. 2010).

8.5   REMARKS AND FUTURE DIRECTIONS

A large amount of work has been done over the years and most intense over the last decade for obtaining bactericidal materials or coatings containing silver which can be released over a period of days or weeks. Despite numerous studies, it is difficult to identify an ideal method that can provide the desired improvement to biopolymeric materials intended for medical applications.

In this chapter, we have demonstrated the advantages of plasma-assisted techniques for the production and modification of biomaterials. The plasma surface modification of biomaterials is an economical and effective method by which biocompatibility and biofunctionality can be achieved while preserving the favorable bulk characteristics of the biomaterial, such as strength and inertness. This provides device manufacturers with a flexible and environmentally friendly process that allows tailoring of the surface properties of the material to suit a specific need.

Despite numerous auspicious results reported in the literature, real-life applications are frequently hindered by a limited understanding of the influence of the process parameters including, among others, the geometry of the reactor, the input energy, and the pressure. The combination of these parameters determines the nature of the reactive species and, ultimately, the surface modifications produced. The combination of plasma and chemical modification adds value to the final product demonstrated by the physicochemical and antibacterial properties. The FTIR–ATR spectra of the treated urinary catheter demonstrate combined effects of conformational changes, molecular orientation, and initial degradation restricted to the amorphous part, chain scission, and reduction of the silver at the treated surfaces. SEM images at micrometric scale suggested that the silver ions are present in the treated surfaces and are uniformly distributed, while the influence of treatment conditions on catheter crystallinity and the reduction of silver in the case of the treated samples was evidenced by XRD measurements. The sensitivity of *Pseudomonas* to the silver-treated catheter-isolated bacterial strain to silver ions attached to the catheter surface was higher for the samples treated first in argon–oxygen and then in $AgNO_3$ solution compared to simple argon plasma treatment followed by wet $AgNO_3$ treatment.

In this chapter, the influence of plasma treatment is demonstrated by the different bactericidal effect as a function of plasma type (simple argon or argon–oxygen combination). The two-step treatment described demonstrates to be an efficient, easy to perform, and low-cost method for avoiding CUTIs related to urethral catheterization.

Further advancement in the areas of immunology, biology, and analytical techniques is necessary for the successful design and implementation of biomaterials.

ACKNOWLEDGMENTS

Research leading to these results has received funding from the PN-II-RU-TE-0123 Project, nr. 28/29.04.2013.

REFERENCES

Aflori, M. 2014. Surface characterization of peritoneal dialysis catheter containing silver nanoparticles. *Rev. Roum. Chim.* 59:523–526.

Aflori, M., Drobota, M., Dimitriu, D.Gh., Stoica, I., Simionescu, B., and Harabagiu, V. 2013. Collagen immobilization on polyethylene terephthalate surface after helium plasma treatment. *Mat. Sci. Eng. B* 178:1303–1310.

Aflori, M., Drobota, M., and Timpu, D. 2008. Studies of amine treatments influence on poly(ethyleneterephthalate) films. *Optoel. Adv. Mat. Rapid Comm.* 2(5):291–295.

Ahearn, D.G., Grace, D.T., and Jennings, M.J. 2000. Effects of hydrogel/silver coating on *in vitro* adhesion to catheters of bacteria associated with UTI. *Curr. Microbiol.* 41:120–125.

Anders, A. 2005. Plasma and ion sources in large area coating: A review. *Surf. Coat. Technol.* 200:1893–1906.

Biederman, H. and Slavínská, D. 2000. Plasma polymer films and their future prospects. *Surf. Coat. Technol.* 125:371–376.

Chu, P.K. 2007. Enhancement of surface properties of biomaterials using plasma based technologies. *Surf. Coat. Technol.* 201:8076–8082.

Darouiche, R.O., Smith, J.A., and Hanna, H. 1999. Efficacy of antimicrobial-impregnated bladder catheters in reducing catheter associated bacteriuria: A prospective, randomised, multicentre clinical trial. *Urology* 54:976–981.

Denes, F.S. and Manolache, S. 2004. Macromolecular plasma-chemistry: An emerging field of polymer science. *Prog. Polym. Sci.* 29:815–885.

Desmet, T., Morent, R., Geyter, N.D., Leys, C., Schacht, E., and Dubruel, P. 2009. Nonthermal plasma technology as a versatile strategy for polymeric biomaterials surface modification: A review. *Biomacromolecules* 10:2351–2378.

Drobota, M., Aflori, M., and Barboiu, V. 2010. Protein immobilization on poly(ethylene tere-phthalate) films modified by plasma and chemical treatments. *Dig. J. Nanomater. Bios.* 5(1):35–42.

Gheorghiu, M., Aflori, M., and Dorohoi, D. 2005. Polyethyleneterephthalate (PET) films interactions with low energy oxygen ions. *J. Optoel. Adv. Mat.* 7(2):841–844.

Gordon, O., Slenters, T.V., Brunetto, P.S., Villaruz, A.F., Sturdevant, D.E., Otto, M., Landmann, R., and Fromm, K.M. 2010. Silver coordination polymers for prevention of implant infection: Thiol interaction, impact on respiratory chain enzymes, and hydroxyl radical induction. *Antimicrob. Agents Chemother.* 54:4208–4218.

Gould, C.V., Umscheid, C.A., Agarwal, R.K., Kuntz, G., and Pegues, D.A. 2010. Guideline for prevention of catheter-associated urinary tract infections. *2009 Infect. Control. Hosp. Epidemiol.* 31(4):319–326.

Grandin, H.M. and Textor, M. 2012. *Intelligent Surfaces in Biotechnology.* John Wiley & Sons, Inc, Hoboken, NJ.

Gristina, R., D'Aloia, E., and Senesi, G.S. 2009. Increasing cell adhesion on plasma depos-ited fluorocarbon coatings by changing the surface topography. *J. Biomed. Mater. Res. B Appl. Biomater.* 88B:139–149.

Hooton, T.M., Bradley, S.F., and Cardenas, D.D. 2010. Diagnosis, prevention, and treatment of catheter-associated urinary tract infection in adults. 2009 International Clinical Practice Guidelines from the Infectious Diseases Society of America. *Clin. Infect. Dis.* 50(5):625–663.

Hsieh, S., Cheng, Y.A., Hsieh, C.W., and Liu, Y.L. 2009. Plasma induced patterning of polydimethylsiloxane surfaces. *Mat. Sci. Eng. B Adv.* 156:18–23.

Huang, N., Leng, Y.X., and Yang, P. 2003. Plasma surface modification of biomaterials applied for cardiovascular devices. *The 30th International Conference on Plasma Science*, Jeju, Korea, *IEEE Conference Record—Abstracts* 439.

Hwang, I.T., Kuk, I.S., Jung, C.H., Choi, J.H., Nho, Y.C., and Lee, Y.M. 2011. Efficient immobilization and patterning of biomolecules on poly(ethylene terephthalate) films functionalized by ion irradiation for biosensor applications. *Appl. Mater. Interfaces* 3:2235–2239.

Inbakumar, S., Morent, R., and De Geyter, N. 2010. Chemical and physical analysis of cotton fabrics plasma-treated with a low pressure DC glow discharge. *Cellulose* 17:417–426.

Jacobs, T., Morent, R., De Geyter, N., Desmet, T., Dubruel, P., and Leys, C. 2010. Effect of humid air exposure between successive helium plasma treatments on PET foils. *Surf. Coat. Tech.* 205:2256–2261.

Johnson, J.R., Roberts, P.L., Olsen, R.J., Moyer, K.A., and Stamm, W.E. 1990. Prevention of catheter associated urinary tract infections with a silver oxide coated urinary catheter. Clinical and microbiologic correlates. *J. Infect. Dis.* 162:1145–1150.

Johnston, E.E. and Ratner, B.D. 1996. Surface characterization of plasma deposited organic thin films. *J. Electron. Spectrosc. Relat. Phenom.* 81:303–317.

Kauling, A.P., Soares, G.V., and Figueroa, C.A. 2009. Polypropylene surface modification by active screen plasma nitriding. *Mat. Sci. Eng. C Mater.* 29:363–366.

Kwok, D.T.K., Tong, L., Yeung, C.Y., Remedios, C.Gd., and Chu, P.K. 2010. Hybrid plasma surface modification and ion implantation of biopolymers. *Surf. Coat. Technol.* 204:2892–2897.

Lai, K.K. and Fontecchio, S.A. 2002. Use of silver hydrogel urinary catheters on the inci-dence of catheter associated urinary tract infections in hospitalised patients. *Am. J. Infect. Control* 30:221–225.

Leroux, F., Perwuelz, A., Campagne, C., and Behary, N. 2006. Atmospheric air–plasma treat-ments of polyester textile structures. *J. Adhes. Sci.Technol.* 20:939–957.

Liedberg, H. and Lundeberg, T. 1989. Silver coating of urinary catheters prevents adherence and growth of *Pseudomonas aeruginosa*. *Urol. Res.* 17:357–358.

Munasinghe, R.L., Yazdani, H., Siddique, M., and Hafeez, W. 2001. Appropriateness of use of indwelling urinary catheters in patients admitted to the medical service. *Infect. Control Hosp. Epidemiol.* 22(10):647–649.

Padawer, D., Pastukh, N., and Nitzan, O. 2015. Catheter-associated candiduria: Risk factors, medical interventions, and antifungal susceptibility. *Am. J. Infect. Control* 43(7):19–22.

Pandiyaraj, K.N., Selvarajan, V., Deshmukh, R.R., and Gao, C. 2008. Adhesive properties of polypropylene (PP) and polyethylene terephthalate (PET) film surfaces treated by DC glow discharge plasma. *Vacuum* 83:332–339.

Patel, H.R.H. and Arya, M. 2000. The urinary catheter: A voiding catastrophic. *Hosp. Med.* 62:148–149.

Raynor, J.E., Capadona, J.R., Collard, D.M., Petrie, T.A., and Garcia, A.J. 2009. Polymer brushes and self-assembled monolayers: Versatile platforms to control cell adhesion to biomaterials (review). *Biointerphases* 4:FA3–FA16.

Riley, D.K., Chassen, D.C., Stevens, L.E., and Burke, J.P. 1995. A large randomised clinical trial of silver impregnated urinary catheter: Lack of efficacy and staphylococcal super-infection. *Am. J. Med.* 98:349–356.

Roth, S., Feichtinger, J., and Hertel, C. 2010. Characterization of *Bacillus subtilis* spore inactivation in low-pressure, low-temperature gas plasma sterilization processes. *J. Appl. Microbiol.* 108:521–531.

Sagripanti, J.L. 1992. Metal based formulations with high microbial activity. *Appl. Environ. Microbiol.* 58:3157–3162.

Saint, S. and Lipsky, B.A. 1999. Preventing catheter related bacteraemia. Should we? Can we? How? *Arch. Intern. Med.* 159:800–808.

Schumm, K. and Lam, T.B. 2008. Types of urethral catheters for management of short-term voiding problems in hospitalised adults. *Cochrane Database Syst. Rev.* 16:CD004013.

Sohbatzadeh, F., Hosseinzadeh Colagar, A., Mirzanejhad, S., and Mahmodi, S.E. 2010. Coli, *P. aeruginosa*, and *B. cereus* bacteria sterilization using afterglow of non-thermal plasma at atmospheric pressure. *Appl. Biochem. Biotechnol.* 160:1978–1984.

Stark, R.P. and Maki, D.G. 1998. Bacteriuria in the catheterised patient: What quantitative level of bacteriuria is relevant? *N. Engl. J. Med.* 311:560–564.

Subbiahdoss, G., Kuijer, R., Grijpma, D.W., Van der Mei, H.C., and Busscher, H.J. 2009. Microbial biofilm growth vs. tissue integration: The race for the surface experimentally studied. *Acta Biomater.* 5:1399–1404.

Sureshkumar, A., Sankar, R., Mandal, M., and Neogi, S. 2010. Effective bacterial inactivation using low temperature radio frequency plasma. *Int. J. Pharm.* 396:17–22.

Tatoulian, M., Arefi khonsari, F., Mabillerouger, I., Amouroux, J., Gheorgiu, M., and Bouchier, D. 1995. Role of helium plasma pretreatment in the stability of the wettability, adhesion, and mechanical properties of ammonia plasma-treated polymers. Application to the Al–polypropylene system. *J. Adhes. Sci. Technol.* 9:923–934.

Terlingen, J.G.A. 1993. *Introduction of Functional Groups at Polymer Surfaces by Glow Discharge Techniques*. University of Twente, Enschede, the Netherlands.

Thomas, L., Valainis, G., and Johnson, J. 2002. A multi site cohort matched trial of an anti-infective urinary catheter. *The Society for Health Care Epidemiology of America 12th Annual Scientific Meeting Final Program*, Salt Lake City, Utah, p. A207.

Vasilev, K., Cook, J., and Griesser, H.J. 2009. Antibacterial surfaces for biomedical devices. *Expert Rev. Med. Devices* 6:553–567.

Vergidis, P. and Patel, R. 2012. Novel approaches to the diagnosis, prevention, and treatment of medical device-associated infections. *Infect. Dis. Clin. North Am. Mar.* 26(1):173–186.

Verleyen, P., De Ridder, D., Van Poppel, H., and Baert, L. 1999. Clinical application of the Bardex IC Foley catheter. *Eur. Urol.* 36:240–246.

Yang, Z., Wang, X., Wang, J., Yao, Y., Sun, H., and Huang, N. 2009. Pulsed-plasma polymeric allylamine thin films. *Plasma Process. Polym.* 6:498–505.

9 Effects of Gamma Irradiation on Polymer Materials Used in Biomedical Application

Cristina-Delia Nechifor

CONTENTS

Abstract

Understanding the influence of gamma exposure on polymers is really important because even a low irradiation dose may induce important modification in the physical structure and physicochemical properties of the polymer, such as chain scission and cross-linking. The chemical resistance, mechanical strength, and physical, chemical, and temperature stability are few properties very important when polymer membranes are gamma irradiated.

Exposure of a polymeric substrate to ionizing radiations is a technique known as radiation processing. It is used in many applications, from industry to medicine, as a substitute for conventional processing techniques based on heating and chemicals addition. The main goal of this technique is to modify the physicochemical properties of the existing material in order to obtain materials with improved or new properties. Current developments in applied radiation chemistry of polymers are focused on radiation cross-linking, curing, controlled degradation, and radiation-induced grafting. Sterilization technique of medical devices in order to remove the microorganisms from the contaminated surfaces is another important approach of gamma ray irradiation.

This section contains a brief introduction about materials used in biomedical fields and three parts which present theoretical concepts about gamma radiation process, the radiation mechanism of polymer materials, and a short review about properties, applications, and the major effects of gamma irradiation on the most common polymers used as biomaterials.

9.1   INTRODUCTION

The use of different materials in the biomedical field has a rapid growth over the last few years, thanks to advances in medicine, pharmaceutical, biochemistry, and materials science. The practical importance of polymers in medical applications is increasingly because of their excellent chemical and physical properties compared to other biocompatible materials (Tsuruta et al. 1993). The polymers, compared to metals, ceramics, and composites, are easily processed, present good tensile strength and elongation, and can be obtained in various shapes and forms at a reasonable cost.

Three types of materials are used in the biomedical field. The first type is represented by the materials which do not come in contact with body tissues (e.g., blood bags, recipients for drugs, tubes, surgical clothing, and equipment). The second class of materials is occasionally in contact with living tissue (e.g., blood oxygenator, dialyzers, patches, catheters, and syringes). The third category includes those materials which are in permanent contact with living tissue such as heart valves, sutures, and artificial organs: heart, lung, kidney, bone grafting, ophthalmological implants, adhesives, and controlled drug delivery systems. This last class of materials is divided into other three categories, depending on how they interact with the biological tissues in bioinert materials, bioactive materials, and biodegradable materials.

For all applications in the biomedical field, the product has to be bio-stable and compatible with the human body and/or with the biological systems. The bio-stability consists of an immune response. This implies that the material will not generate toxic products, will not be pyrogenic, inflammatory, cytotoxic, antigenic, oncogenic, or not biodegradable (Park and Roderic 1992), it can be sterilized and it will not lose its properties under physical and chemical external conditions.

There are some important properties needed to be accomplished when a material is chosen for a specific biomedical application. The material selection process is made according to various requirements listed below. The physical properties are in the sense of the dimensions, size, weight, transparency/opacity, water absorption, and wear resistance, demanded for the product. The mechanical properties such as tensile strength, tensile elongation, tensile modulus, impact resistance, and flexural modulus are necessary to assure a good toughness. The important thermal properties of a polymer are melting point, processing temperatures, heat deflection, glass transition temperature, and thermal conductivity. The electrical properties such as conductivity and dielectric strength are also very important. The chemical resistance to various types of oils, greases, processing aids, disinfectant, bleaches, and the resistance to repetitive sterilization methods are mandatory aspects in the selection process (McKeen 2014).

Sterilization is a process required in all the biomedical applications which is used to eliminate almost all microorganisms such as fungi, bacteria, viruses, and spore forms. A common method of sterilization is by using gamma rays. This method is generally used for the sterilization of medical devices, such as syringes, needles, cannulas, etc. (Da Silva Aquino 2012). Although there are many different sterilization methods, the gamma ray sterilization has some considerable advantages which include superior assurance on the product sterility; it leaves no residue behind and low-temperature process.

When gamma radiation is used as a technique for the sterilization of materials from the biomedical field, it will not only kill the microorganisms, but it will also affect the material properties. Understanding the effects of gamma exposure on polymers is really important because even a low irradiation dose may induce important modification in physicochemical proportion of the polymer such as chain scission and cross-linking (Schnabel 1981).

9.2   GAMMA RADIATION: OVERVIEW

Gamma rays are ionizing radiations which have a very short wavelength of less than one-tenth of nanometers (Attix and Roesch 1968), being the most energetic form of electromagnetic radiation. It is emitted by the radioactive decay representing the spontaneously disintegration of the unstable isotopes (Masefield 2004). Among the most popular gamma ray sources are Cobalt-60, Cesium-137, Iodine-131, Americium-241, Radium-226, and Radon-222. Of all, the most proper gamma radiation source for radiation processing is Cobalt-60 because of the reasonably high energy and relatively long half-life (5.27 years) (Chmielewski and Haji-Saeid 2004).

When gamma radiation interacts with matter, three different processes can take place depending on photon energy (Attix and Roesch 1968). These processes are described below:

Photoelectric effect consists of electron ejection from the K shell of an atom by the photon. It takes place if the photon energy is less than 0.1 MeV or if matter has a high atomic number. The electromagnetic radiation is absorbed and the ejected electron, having the energy of the photon less than the binding energy of the electron in atom, travels and interacts with other atoms.

Compton effect (scattering) appears if the photon energy is higher than 0.1–10 MeV. In this process, a high-energy photon and a free or loosely bound electron will collide leading to a consequent loss of its energy. The electron will absorb the energy of the photon and it will recoil. The gamma radiation goes off in another direction.

Pair production consists of simultaneous creation of a positron (positive electron) and an electron as a result of interaction between electromagnetic radiation and the field of an atomic nucleus. The energy of photons, for pair production, should be more than 1.02 MeV. If the electron and positron will recombine, two identical gamma rays are produced, named annihilation radiation with energy of 0.51 MeV.

When a material is deliberately exposed to gamma radiation, usually it pursues to preserve, modify, or improve its properties (Chmielewski and Haji-Saeid 2004). So, for each type of material, an exact amount of radiation energy is required to obtain the desired effect, which can be controlled by the adjustment of the exposure time. The dose needed to reach a desired effect is known as process dose. It is determined through experiments which involve establishing the dose–effect relationship for the desired result (Attix and Roesch 1968).

Generally two dose restrictions can be established: the lower dose limit, which indicates the minimum dose that is necessary to get the desired effect, and the upper dose limit, which is fixed to ensure that radiation will not unfavorably affect the functional quality of the product. The ratio between the higher dose limit and the inferior dose limit is known as dose limit ratio.

The measure of absorbed radiation energy, named radiation-absorbed dose, is the energy deposited by ionizing radiation per unit mass. It is measured as joules per kilogram and is represented by the equivalent SI unit, gray (Gy). The quantity of radiation absorbed per unit time is known as dose rate. It is measured in Gy/s and depends on the radiation source activity and on the irradiation geometry. The power of a radiation source, which is defined as the number of decays of radioactive nuclides per second, is known as radioactivity level. The SI unit is becquerel (Bq), but it can also be measured in curie (Ci) (Attix and Roesch 1968).

As it can be noticed, the desired effect can be obtained for a certain quantity (mass or volume) of material processed per unit time and is dependent on the radioactivity level of the radiation source, material density, and the absorbed dose.

9.3   RADIATION MECHANISM ON POLYMER MATERIALS

Polymers are known as a versatile class of materials which are frequently used in biomedical applications. The ability to be casted or prepared into very complex shapes is the most significant property of polymers (David 2002). Polymers are also easily manufactured, the costs of these materials processing being relatively low (Hegemann et al. 2003, Fare et al. 2005). They have an extensive area of properties, such as elasticity, conductivity, strength, degradability, and biocompatibility (Goddard and Hotchkiss 2007), which makes them available for a large range of biomedical applications. These are few reasons why scientists continue to develop new polymers, original polymerization processes, and also different physical and chemical methods to modify the existing polymers.

A physical method used to modify or improve the properties of polymers is irradiation process. This process occurs by placing the material in the surroundings of a radiation source for a time interval. The medical and healthcare fields are major axes in which modified polymers are used. These materials found applications as implantable materials, controlled release drug systems, wound dressings (Hegemann et al. 2003), and sterilization of medical devices (Hoyle and Kinstle 1990). Other areas include the textile, mining, and construction industries (Clough and Shalaby 1991).

Utilization of high-energy radiation in polymer processing for biomedical application is restricted by the impossibility to prevent unwanted material property changes. This unwanted effect can be avoided if the technique involves small chemical changes applied to polymers (Gürsel et al. 2008). The role of radiation process is certainly very significant when the goal is to obtain new systems with considerable physical and biological benefits, but also is important to elucidate the effects of high-energy radiation on physicochemical properties of a specific polymer material.

Gamma radiation interacts with the substance in the primary stages by the mechanism of excitation of its atoms and molecules converting them into ions with different signs, this mechanism being named ionization. The excited and ionized species are the initial chemical reactants for graft polymerization (Drobny 2003). The resulting species can further react to give free radicals. These species, having an unpaired electron, which results from cleavage of a chemical bond, may rapidly cross-link the polymer chains, thereby increasing the molecular weight, hardness, and wear resistance (Dole 1972). Free radicals can also cause monomers to polymerize, polymers to degrade, and when used in mixtures, they can cause monomers to graft to polymers (Drobny 2003). Permanent damage in the polymer structure, such as breaking the chains of macromolecules (Schmitz 1966) and the reduction of average molecular weight, can be caused by the displacement of atoms from polymer chains when excited and ionized species collide with the targeted atoms. Changes in the physical properties of the material, such as discoloration, the emission of gas (Drobny 2003), soaking, and modification of the melting temperature (Abraham et al. 1997), were also reported.

These changes in properties may have both detrimental and beneficial consequences in determining the end uses of the polymer. It is useful in the sense that it can cause cross-linking and grafting on the surface of the polymers, but on the other hand, it may cause chain scission leading to considerable damaging of the polymer. The qualities of these processes are strongly dependent not only on the chemical structure of the polymer (Gürsel et al. 2008), but also on the physical parameters (e.g., type of the radiation, temperature of the irradiation, presence or absence of oxygen, and irradiation doses) (Swallow 1960). The presence of oxygen in the environment leads to rapid oxidation reactions, which are reflected in the physical property changes. Different sites onto the macromolecule will be preferentially oxidized because of the movement of radicals inside the polymer, which takes place in order to obtain the most stable species (Gopal et al. 2007).

Some chemical bonds and groups such as COOH, C–X, where X = halogen, SO_2^-, NH_2, and C=C can be more sensitive to high-energy radiation and others are known to have radiation resistance, for example the aromatic groups (Reichmanis et al. 1993). The stability of a polymer to radiation is strongly dependent on its chemical structure. The polymers containing aliphatic groups (ethers and alcohols) showed to be least resistant to radiation relative to polymers containing aromatic groups (Roy et al. 2012). A significant increase of stability to radiation was demonstrated for polymers containing phenyl groups (e.g., polystyrene [PS], polyarylene sulfones), as well as for aromatic polymers. The aromatic substituents, biphenyl, and phenolic moieties revealed a similar effect to be responsible with a pronounced radiation resistance of polymer materials compared to other aromatic groups (Dole 1973).

Although polymers containing aromatic groups are resistant to ionizing radiations, these materials are still very expensive. For this purpose, specific chemical agents, named radiation stabilizers, were developed mainly for the stabilization of general polymers. Stabilizers and additives are usually used in small quantities (<1%) in polymer processing. The stabilizers, role is to control and diminish the danger of deterioration of properties (Da Silva Aquino 2012).

9.4 EFFECTS OF GAMMA IRRADIATION ON PHYSICO-CHEMICAL PROPERTIES OF POLYMER MATERIALS USED IN BIOMEDICAL APPLICATIONS

Many of the synthesized polymers can be used in different applications but only a few of them are used in the biomedical field. In this section, the most common polymers used as biomaterials are discussed. General information about their properties and applications is given and the major effects of gamma irradiation of these materials are also listed.

Polyvinylchloride (PVC) is an amorphous, rigid polymer with high-melt viscosity. Membranes and films from PVC are used for blood and solution storage bags and for surgical packaging. Tubes of PVC are ordinarily used in intravenous (IV) administration, dialysis devices, catheters, and cannulas (Da Silva Aquino 2012). Thermal stabilizers, lubricants, and some plasticizers could be incorporated to prevent thermal degradation and adhesion.

Tensile strength, elongation, number of scissions per chain, viscosity–average molecular weight, and ultraviolet (UV)-visible absorbance are few properties affected by the exposure of PVC to gamma radiation. Main chain scission and cross-linking effects change the properties of PVC depending on the irradiation conditions, the used plasticizer type, and the dose rate. It was noticed that the addition of Tinuvin P® significantly decreased the number of scissions per chain (Vinhas et al. 2004).

When gamma radiation interacts with PVC radicals deriving from CCl, bond scission reactions appear. These chlorine radicals form HCl which acts as a catalyst, they can recombine with each other forming networks (Baccaro et al. 2003), or if air is present, the polymeric radicals react with oxygen producing the peroxyl macroradical.

Polyethylene (PE) is a polymer which exists in five major classes as a consequence of different synthesizing conditions: (1) high-density polyethylene (HDPE), (2) low-density polyethylene (LDPE), (3) linear low-density polyethylene (LLDPE), (4) very-low-density polyethylene (VLDPE), and (5) ultra-high-molecular-weight polyethylene (UHMWPE). HDPE is used for pharmaceutical bottles and caps. LDPE is found in flexible container applications, nonwoven-disposable, and laminated foil. LLDPE is frequently used in pouches and bags and VLDPE is used in extruded tubes. UHMWPE has found application as orthopedic implant fabrications (Da Silva Aquino 2012).

Increase in percent of crystallinity, density, elastic modulus, and a decrease in elongation to failure are the consequence of oxidative degradation caused by gamma exposure time of UHMWPE (Costa and Bracco 2004). For LDPE films, irradiated with gamma radiation, in the dose range varied from 20 to 400 kGy, the induced changes in the chemical structure and dielectric properties were investigated. Results indicated a small variation in crystallinity, which could be increased or decreased depending on the relative importance of the structural and chemical changes (Abdel Moez et al. 2012).

Polypropylene (PP) is one of the most currently used polymers in the manufacturing of medical devices. It has high transparency, good mechanical properties, and low production cost. To improve its properties, additives such as antioxidants, light stabilizers, nucleating agents, lubricants, mold release agents, antiblock, and slip agents can be added. PP is usually used to make disposable hypothermic syringes, blood oxygenator membranes, packaging for devices, solutions, and drugs, suture, or artificial vascular grafts (Da Silva Aquino 2012).

Changes in polymer properties due to oxidative degradations were observed when PP was gamma irradiated in the presence of air. Some studies relate the decrease of the melting temperature, viscosity, and mechanical properties. Increases in crystallinity were noticed and explained due to the chain scissions and recrystallization of shorter chains which formed new crystallites. The cross-linking effect was also revealed depending on the irradiation dose (Kushal and Praveen 2003).

Polymethylmethacrylate (PMMA) is one of the most biocompatible polymers. It is an amorphous material with high light transparency (92% transmission) and relatively high refractive index (1.49). PMMA is used mainly in medical applications for blood pumps and reservoirs, membranes for blood dialyzer, contact lenses and implantable ocular lenses, and prostheses (Da Silva Aquino 2012).

Gamma irradiation of PMMA causes main scission and hydrogen abstraction from an α-methyl or methylene group, which results in radicals appearance, responsible for changes in physical properties (Schnabel 1981). Depending on irradiation conditions, oxidative products can be generated inducing further degradation of PMMA.

PS is an aromatic polymer available in three grades: unmodified general-purpose PS (GPPS) commonly used in tissue culture flasks, roller bottles, vacuum canisters, and filterware; high-impact PS (HIPS), and PS foam. It is chemically inert, being resistant to acids and bases, but is easily dissolved by chlorinated solvents and aromatic hydrocarbon solvents. The aromatic rings from its structure provide a protective action toward radiation effects, being the most radiation stable of the common polymers, large doses being required to bring significant effects. Although gamma radiation modifies the surface chemical properties of PS producing surface >C=O and C–O-containing functional groups, it also causes oxidation to depths >10 nm. Surface oxidation regards pseudo-first-order kinetics, the extent of interior oxidation being linear with dose (Onyiriuka et al. 1990).

Polyesters (PES) are a category of polymers that contain the ester functional group in their main chain. As a specific material, it most commonly refers to polyethylene terephthalate (PET) which is frequently found in medical applications, such as artificial vascular graft, sutures, and meshes. As a material for medical devices, PET is suitable for radiation sterilization using gamma radiations, although discoloration occurs even at lower doses. Cross-linking and oxidation are the major effects of gamma irradiation. The PES begins to deteriorate if the radiation dose is over 50 kGy. Irradiation causes the fractional degradation of the fiber from surface and the chain scission resulting in low-molecular weight chains (Bobadillu Sánchez et al. 2009).

Polyamides (PA) are known as nylons and used as packaging films, catheters, and sutures. Nylons are hygroscopic and lose their strength *in vivo* when implanted (Da Silva Aquino 2012). The effects of gamma radiation were studied on different types of nylon. When nylon 6–12 was exposed to low dose, a cross-link mechanism in the amorphous zone was considered, while high radiation doses induced chain scission, which lead to a decrease in the fusion temperature (Menchaca et al. 2012). Oligomers formation, chain reorganization, and creation of new hydrogen bonds leading to the formation of new crystalline domains were evidenced for long-time gamma radiation exposure. Also, the presence of oxygen substantially increases the effects of radiation (Kumar Gupta et al. 2014).

From the *fluorocarbon polymer* class, known to be characterized by a high resistance to solvents, acids, and bases, the most familiar in biomedical applications is polytetrafluoroethylene (PTFE) (Da Silva Aquino 2012). The category of fluorocarbon polymers also includes polyvinylfluoride (PVF), polytrifluorochloroethylene (PTFCE), fluorinated ethylene propylene (FEP), polyethylenetetrafluoroethylene (ETFE), and others.

PTFE is usually used for implantable devices and in drug delivery. PTFE is very resistant to heat and to chemical exposure, but it seems to be extremely sensitive to radiation, especially when irradiation is in the presence of air. When low-crystalline PTFE was gamma irradiated, it showed significant changes in modulus and density than highly crystalline material. Irradiation reduced the tensile strength of the PTFE mainly by reducing the strain-hardening contribution to the strength. The highly crystalline PTFE showed little strain hardening initially, it lost all ductility, but its tensile strength was little affected (McLaren 1965). The results showed that after chain scission, occurring during irradiation, the cross-linking takes place increasing the crystallinity and making the polymer more rigid and brittle.

Rubbers such as silicone, natural, and synthetic ones have been used for the fabrication of implants from a long time. Natural rubber was found to be compatible with blood in its pure form; new synthetic rubbers were developed as substitutes for the natural rubber, rarely used as implants. One polymer developed especially for medical use is silicone rubber. Silicone rubber is generally nonreactive, stable, and resistant to extreme environments and temperatures. Liquid silicone rubber is manufactured for biomedical applications such as syringe pistons, closure for dispensing system, gaskets for IV low regulator, respiratory masks, and implantable chambers for IV administration.

It was noticed that the hardness and modulus of rubber increase with gamma radiation dose, being proportionally with the cross-link density of the rubber. Damage is reduced in the absence of oxygen, but is much accelerated if irradiation takes place under conditions of stress. For doses exceeding 100 kGy, degradation of rubbers becomes independent of atmosphere and dose rate (Şen et al. 2003). Gamma radiation induces changes in the molecular architecture of rubbers resulting in an increase in molecular weight (Da Silva Aquino 2012).

Polyurethanes (PU) are usually thermosetting polymers, which do not melt when they are heated; so, they are widely used for coat implants. The polyurethane rubber is quite strong and it has good resistance to oil and chemicals (Da Silva Aquino 2012).

Irradiating polyurethane, two competitive processes such as degradation and cross-linking take place. It was evidenced that when irradiation occurs in air, the thermal stability increases compared to irradiation in water (Pârvu et al. 2010). Detailed discussions about the effects of gamma radiation on PU are presented in the following sections of this chapter.

Polyvinyl alcohol (PVA) is a transparent optical polymer widely used, alone or in blends, to obtain different biomedical materials (Feng et al. 2005, Zhang et al. 2007), because it can be easily obtained and processed at a low cost (Tagaya et al. 2006). The effects of gamma radiation on the induced birefringence of PVA-stretched foils with comparable thickness are discussed below.

The birefringence of polymers is the result of the asymmetry of their molecular structure (Bhattacharya et al. 2009). A polymer material is usually characterized by an intrinsic optical anisotropy (Barbu and Bratu 1997) due to the random orientation of the different macromolecular chains. If the polymer chains are oriented in a certain direction, the optical anisotropies are no longer canceling each other (Zhang et al. 2007) and the material becomes birefringent. Analyzing the birefringence of the stretched polymer foils offers important information about the order degree of the polymer chains (Gürsel et al. 2008).

When PVA (hydrolyzed 99%, with average Mw = 22,000) foils, with comparable thickness (100 µm), obtained from 10 wt.% PVA solutions, were exposed to gamma source ^{137}Cs, for 15, 45, 90, and 360 min in the presence of air, at room temperature, an increase of birefringence versus the stretching ratio was evidenced. The stretching degree and the anisotropy of the etired foils (stretching the foils at 45°C) were influenced by the physico-chemical processes induced by gamma irradiation of the polymer matrix. The birefringence was found to be dependent on the exposure time and the stretching ratio.

The birefringence of the stretched polymeric foils was expressed as the difference between the two main refractive indices, which are usually different: the extraordinary refractive index of the film for linearly polarized radiation having the electric field intensity parallel to the stretching direction, and the ordinary refractive index of the film for linearly polarized radiation having the electric field intensity perpendicular to the stretching direction (Dole 1972). The experimental device used for birefringence measurement consists of two identical cross-seated polarizers. When an anisotropic layer (PVA-stretched foil) is introduced between the crossed polarizers, the polarization state will change. In order to estimate the birefringence, the pathway introduced by the film is compensated by a Babinet Compensator (BC) which is placed into the device between the thin film and the analyzer. Initially the BC was standardized in a monochromatic radiation, the yellow radiation of a Na lamp, $\lambda = 589.3$ nm.

The stretching ratio is represented by the fraction between the length of the large semi-axis and the length of the small semi-axis of the ellipse which results, after stretching, from a control circle drawn upon the film.

The dependence of the induced birefringence on the stretching ratio for the non-exposed PVA foils is given in Figure 9.1.

From Figure 9.1, it results that the tendency of the birefringence increases linearly with the stretching degree. It can be seen that the increase tendency is divided into two regions. The limit of these regions can be established, according to the slope of the trendlines, at a value of the stretching degree about 1.25. A bigger slope of the trendline suggests an appropriate alignment of the polymer chains. The existence of the limit in the stretching degree suggests the existence of a point at which the alignment mechanism is changing (Nechifor et al. 2014).

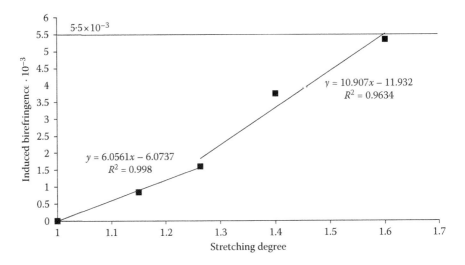

FIGURE 9.1   Dependence of the induced birefringence on the stretching ratio for nonexposed PVA foils.

The dependence of the induced birefringence for the gamma-exposed PVA foils versus the stretching ratio is presented in Figure 9.2.

The graphs from Figure 9.2 illustrate that the induced birefringence of gamma-exposed samples increases with the exposure time. At the same ratio of stretching, the alignment degree of polymer chains increases with gamma exposure time, but it does not overcome the nonexposed foils birefringence. The induced birefringence

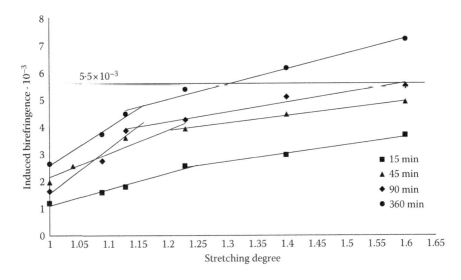

FIGURE 9.2   Dependence of the induced birefringence on the stretching ratio for PVA foils exposed to gamma radiations.

is lower after gamma irradiation, which indicates the existence of a lower degree of alignment of the polymer chains. Only for the PVA foil expose at 360 min of gamma ray, the birefringence rises above the nonexposed foil. If the polymer chain cleavage is followed by internal cross-linking reactions for long-term gamma-irradiated PVA foils, which can induce restriction in chain mobility, an inferior alignment can be observed through stretching (Qi et al. 2004).

As it can be observed, the slope of trendline of the dependence between induced birefringence and stretching ratio is changing as a function of the exposure time. The slope of the gamma-exposed foils decreases after a specific stretching degree compared to nonexposed PVA foils. A decrease of slope indicates a saturation of the birefringence at a stretching degree over that value. The slope becomes smaller due to the orientation relaxation and limitation of chain alignment (Qi et al. 2004, Nechifor et al. 2014).

The stretching ratio at which the slope of trendline is changing decreases linearly with radiation exposure time comparatively with the nonexposed PVA foil. This phenomenon was associated with the photooxidation and photo-degradation processes (Kwak and Yang 2010) which take place. These processes are increasing the mobility of polymeric chains involving a faster relaxation time of chain orientation and an easier alignment.

The appearance of photooxidation and photo-degradation processes was evidenced evaluating the contact angles of the samples for three test liquids: water, formamide, and glycerol using the sessile drop technique. All the contact angles of liquids decrease with gamma irradiation time. The surface polarity and surface free energy of hydration ΔG_w were evaluated.

The ΔG_w values were obtained from Equation 9.1 (Şakar-Deliormanli 2013) by using the total surface tension of water as 72.8 mN/m (Jasper 1972) and the contact angle of water for the studied films.

$$\Delta G_w = -\gamma_l \cdot (1 + \cos\theta_w) = -Wa \qquad (9.1)$$

A value for $\Delta G_w < -113$ mJ/m indicates that the explored material can be considered hydrophilic, while a value for $\Delta G_w > -113$ mJ/m reveals that it should be considered hydrophobic.

The surface polarity is plotted in Figure 9.3 and the surface free energy of hydration in Figure 9.4, both being expressed as a function of gamma exposure time.

During the gamma irradiation, the surface polarity of PVA foils increases. According to Figure 9.4, the value of surface free energy of hydration is reaching the limit between hydrophilicity and hydrophobicity after 45 min of radiation exposure. The surface of PVA foils becomes hydrophilic. Thus, the results from the surface polarity of PVA foils and the surface free energy reveal that the studied samples possess an increased wettability, which is enhanced as the irradiation time is smaller. For a long term of radiation exposure, the surface polarity increases suggesting the improvement of surface bondability. The improvement of the surface bondability can be explained through many mechanisms including increasing the wettability, inducing hydrogen bonding, and the formation of polymeric scission products, which

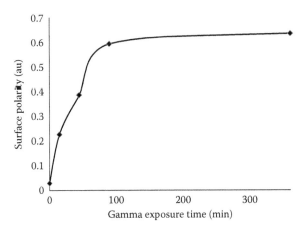

FIGURE 9.3   Surface polarity of PVA foils versus gamma exposure time.

promote interfacial flow, interdiffusion, and polar interactions (Barbu and Bratu 1997, Miyagawa et al. 2014).

Blending *PVA* and *polyhydroxyurethane (PHU)* in different concentrations (12%, 25%, 50%, and 70%) leads to membranes with improved hydrophilic character. These films were found to have changes in surface energy, resilience, and initial elastic module based on PHU concentration (Nechifor et al. 2009a). PHU is an aliphatic polymer from the polyurethane class. It is an elastic polymer used in electronic and electrical industries for packing fragile objects (Ciobanu and Koncsag 2006), subaquatic cables, in constructions, in medicine for catheters, and tubes.

When PVA/PHU blends were exposed to gamma ray using ^{137}Cs source, for 2, 4, and 10 h, at a distance of about 20 cm, in air, important modifications in the

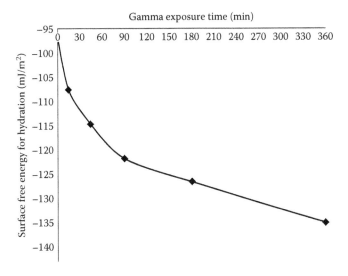

FIGURE 9.4   Surface free energy of hydration of PVA foils versus gamma exposure time.

absorbencies of the attenuated total reflection–Fourier transform infrared (ATR–FTIR) spectra were reported. The noticed changes in frequent vibrations were corresponding to asymmetric stretching vibrations of CH_2 groups, stretching and bending vibrations from amide II, amide III bands, and stretching vibrations of the carbonyl groups (Nechifor et al. 2009b).

The existences of oxidative processes were evidenced by analyzing the changes induced in absorption signals of carbonyl groups (1330 cm^{-1}) and amide II (1550 cm^{-1}) from ATR spectra. To better assess the rate of photo-bleaching and photo-degradation of polymer material, the ratios $R(C)$ and $R(A)$ function of gamma irradiation time were evaluated. $R(C) = A_{carbonyl}/A_{hydroxyl}$ is the ratio between the peak area corresponding to the stretching vibrations of the carbonyl group and the peak area corresponding to the stretching vibrations of the hydroxyl and $R(A) = A_{AmideII}/A_{hydroxyl}$ is the ratio between peak area corresponding to amide II group and hydroxyl corresponding to the band area. The average rate of change in $R(C)$ and $R(A)$ was evaluated from the path of the line obtained by plotting $R(C)$, respectively $R(A)$, versus gamma exposure time. The average rate of change seems to rapidly decrease for the first 2 h of gamma exposure (see Figure 9.5). An increase of rate was observed in both cases in the range of 4–10 h of gamma exposure (see Figure 9.6).

The ratios $R(C)$ and $R(A)$ present a rapid decrease after 2 h of gamma irradiation, indicating that a number of carbonyl and amide II groups of the polymer surface are disappearing. When molecules absorb radiation, inter- and intramolecular rotation and translation processes may occur, leading to changes in surface configuration and change of the $R(C)$ and $R(A)$ ratios.

As it can be observed, from Figure 9.6, both $R(C)$ and $R(A)$ rates of change are increasing with the exposure time. Carbonyl groups can be usually used as an indicator of chemical changes occurring during the oxidative degradation. So, increasing the rate of change in the absorption band of carbonyl region in the course of gamma irradiation can be associated with the generation of oxidized groups (Nechifor et al. 2009b). The observed changes in the $R(A)$ ratio may be attributed to the increase of the frequency vibrations of the amide II group. These can be obtained if the mobility

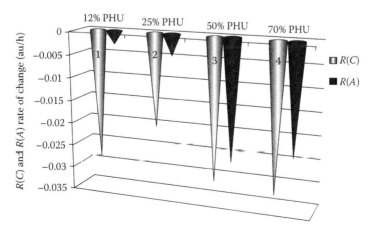

FIGURE 9.5   Average rate of change of $R(C)$ in the first 2 h of gamma exposure.

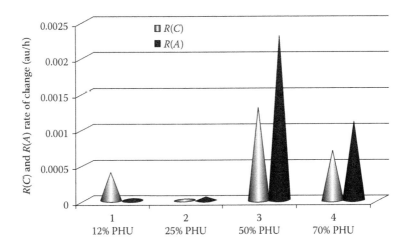

FIGURE 9.6   Average rate of change of $R(A)$ ratios in the range of 4–10 h of gamma exposure.

of functional groups increases as a result of fragmentation of the polymer chains when the dose of gamma radiation grows (Roy et al. 2012). The average rate of change of both ratios seems to be strongly dependent on the PHU concentration of the samples. It seems that the presence of urethane groups makes the blend PHU–PVA more sensitive to gamma radiation (Nechifor et al. 2009b).

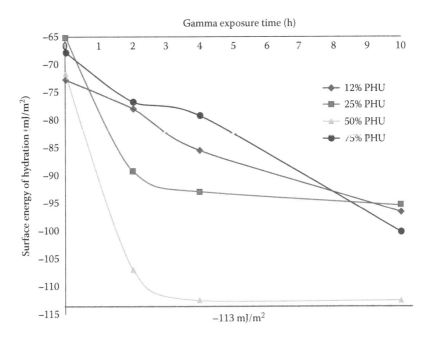

FIGURE 9.7   Dependencies of surface free energy of hydration on gamma exposure time.

These effects are also demonstrated in the surface free energy of hydration (ΔG_w) analysis. The balance between hydrophilicity and hydrophobicity on the sample surface can be analyzed from the dependence of surface free energy of hydration versus gamma exposure time.

The surface free energy of the studied samples possesses an increased wettability, which was enhanced as the gamma irradiation time was longer. According to Figure 9.7, for the sample containing 50% PHU, after 4 h of gamma exposure, the values of ΔG_w almost reached the limit between hydrophilicity and hydrophobicity. Also, as the treatment duration was increased, the difference between the surface free energy of hydration values of the samples becomes smaller.

9.5 CONCLUSIONS

This chapter contains a review of the effects of gamma irradiation on the most common polymer materials used in biomedical application. The discussions were focused on a well-known polymer: PVA and on some blends PVA/PHU in different proportions. Considerations from this chapter still can be completed as long as many recent developments could not be incorporated.

The biomedical field is a very diverse domain, which offers, due to its multidisciplinary nature, different directions of research. It is in a continuous progress because of the new materials developed by the polymer scientists. A trend in this field is to obtain the so-called "smart polymers." These polymers found diverse applications because they can sustain reversible physical or chemical changes in response to the external changes in the environmental conditions (e.g., temperature, pH, light, magnetic or electric field, ionic factors, biological molecules, etc.).

REFERENCES

Abdel Moez, A., Aly, S.S., and Elshaer, Y.H. 2012. Effect of gamma radiation on low density polyethylene (LDPE) films: Optical, dielectric and FTIR studies. *Spectrochim. Acta Part A: Mol. Biomol. Spectrosc.* 93:203–207.

Abraham, G.A., Frontini, P.M., and Cuadrado, T.R. 1997. Physical and mechanical behavior of sterilized biomedical segmented polyurethanes. *J. Appl. Polym. Sci.* 65(6):1193–1203.

Attix, F.H. and Roesch, W.C. 1968. *Radiation Dosimetry.* Academic Press, New York.

Baccaro, S., Brunella, V., Cecilia, A., and Costa, L. 2003. γ Irradiation of poly (vinyl chloride) for medical applications. *Nucl. Instrum. Methods Phys. Res. Sect. B: Beam Interact. Mater. At.* 208:195–198.

Barbu, A. and Bratu, I. 1997. Structural investigations of UV-irradiated packaging polymeric foils. *J. Mol. Struct.* 410(1):229–231.

Bhattacharya, A., Rawlins, J.W., and Ray, P. 2009. *Polymer Grafting and Cross Linking.* John Wiley & Sons, Inc., New Jersey.

Bobadilla-Sánchez, E.A., Martínez-Barrera, G., Brostow, W., and Datashvili, T. 2009. Effects of polyester fibers and gamma irradiation on mechanical properties of polymer concrete containing $CaCO_3$ and silica sand. *eXPRESS Polym. Lett.* 3(10):615–620.

Chmielewski, A.G. and Haji-Saeid, M. 2004. Radiation technologies: Past, present and future. *Radiat. Phys. Chem.* 71:17–21.

Ciobanu, C. and Koncsag, C. 2006. *Sustainable Composite Materials to Improve Quality of Life.* Pim, Iaşi, Romania.

Clough, R. and Shalaby, S.W. 1991. Radiation effects on polymers. In *ACS Symposium Series 475*. ACS, Washington, DC.

Costa, L. and Bracco, P. 2004. Mechanisms of crosslinking and oxidative degradation of UHMWPE. In *The UHMWPE Handbook*, pp. 235–250, ed. S. Kurtz, Academic Press, Boston.

Da Silva Aquino, K.A. 2012. Sterilization by gamma irradiation. In *Gamma Radiation*, pp. 171–206, ed. F. Adrovic, InTech, Rijeka, Croatia.

David, I.B. 2002. *An Introduction to Polymer Physics*. Cambridge University Press, New York.

Dole, M. 1972. Fundamental processes and theory. In *The Radiation Chemistry of Macromolecules*, vol. 1: pp. 25–92, ed. M. Dole, Academic Press, New York.

Dole, M. 1973. Radiation chemistry of substituted vinyl polymers. In *The Radiation Chemistry of Macromolecules*, vol. 2: pp. 4–116, ed. M. Dole, Academic Press, New York.

Drobny, J.G. 2003. *Radiation Technology for Polymers*. CRC Press, New York.

Fare, S., Valtulina, V., and Petrini, P. 2005. *In vitro* interaction of human fibroblasts and platelets with a shape-memory polyurethane. *J. Biomed. Mater. Res. Part A* 73:1–11.

Feng, W., Wang, J., and Wu, Q. 2005. Preparation and conductivity of PVA films composited with decatungstomolybdovanadogermanic heteropoly acid. *Mater. Chem. Phys.* 93(1):31–34.

Goddard, J. and Hotchkiss, J. 2007. Polymer surface modification for the attachment of bioactive compounds. *Prog. Polym. Sci.* 32:698–725.

Gopal, R., Zuwei, M., Kaur, S., and Ramakrishna, S. 2007. Surface modification and application of functionalized polymer nanofibres. In *Molecular Building Blocks for Nanotechnology: From Diamondoids to Nanoscale Materials and Applications*, eds. A.G. Mansoori, T.F. George, F. Assoufid, and G. Zhang, 71–91. Springer Science + Business Media, LLC, New York.

Gürsel, S.A., Gubler, L., Gupta, B., and Scherer, G.G. 2008. Radiation grafted membranes. *Adv. Polym. Sci.* 215:157–217.

Hegemann, D., Brunner, H., and Oehr, C. 2003. Plasma treatment of polymers for surface and adhesion improvement. *Nucl. Instrum. Methods Phys. Res. Sect. B: Beam Interact. Mater. At.* 208:281–286.

Hoyle, C.E. and Kinstle, J.F. 1990. Radiation curing of polymeric materials. In *ACS Symposium Series 417*. ACS, Washington, DC.

Jasper, J.J. 1972. The surface tension of pure liquid compounds. *J. Phys. Chem. Ref. Data* 1(4):841–1009.

Kumar Gupta, S., Singh, P., and Kumar, R. 2014. Modifications induced by gamma irradiation upon structural, optical and chemical properties of polyamide nylon-6, 6 polymers. *Radiat. Eff. Defects Solids* 169(8):679–685.

Kushal, S. and Praveen, K. 2003. Influence of gamma-irradiation on structural and mechanical properties of polypropylene yarn. *J. Appl. Polym. Sci.* 55(6):857–863.

Kwak, C.H. and Yang, H.R. 2010. Determinations of optical field induced nonlinearities. In *Azo Dye Doped Polymer Film, Polymer Thin Films*: pp. 18–36, ed. A.A. Hashim, InTech, Rijeka, Croatia.

Masefield, J. 2004. Reflections on the evolution and current status of the radiation industry. *Radiat. Phys. Chem.* 71:8–15.

McKeen, L.W. 2014. Plastics used in medical devices. In *Handbook of Polymer Applications in Medicine and Medical Devices*: pp. 34–47, eds. K. Modjarrad and S. Ebnesajjad, Elsevier Inc., Oxford.

McLaren, K.G. 1965. Dynamic mechanical studies of irradiation effects in polytetrafluoroethylene. *Brit. J. Appl. Phys.* 16(2):185–193.

Menchaca, C., Demesa, G., and Santiaguillo, A. 2012. Gamma irradiation effect on nylon 6–12 modification under argon atmosphere. *J. Mater. Sci. Eng.: B* 2(4):247–254.

Miyagawa, M., Akai, N., and Nakata, M. 2014. UV-light induced conformational changes of 3-chlorosalicylic acid in low-temperature argon matrices. *J. Mol. Struct.* 1058:142–148.

Nechifor, C.D., Ciobanu, C.L., Dorohoi, D.O., and Ciobanu, C. 2009a. Polymeric films properties of poly (vinyl alcohol) and poly (hydroxy urethane) in different concentrations. *Univ. Politech. Buchar. Sci. Bull. Ser. A* 71(1):97–106.

Nechifor, C.D., Dorohoi, D.O., and Ciobanu, C. 2009b. The influence of gamma radiations on physico-chemical properties of some polymer membranes. *Rom. J. Phys.* 54(3–4):349–359.

Nechifor, C.D., Zelinschi, C.B., and Dorohoi, D.O. 2014. Influence of UV and gamma radiations on the induced birefringence of stretched poly (vinyl)alcohol foils. *J. Mol. Struct.* 1062:179–184.

Onyiriuka, E.C., Hersh, L.S., and Hertl, W. 1990. Surface modification of polystyrene by gamma-radiation. *Appl. Spectrosc.* 44(5):808–811.

Park, J.B. and Roderic, S. 1992. *Biomaterials: An Introduction*, ed. J. Park, Plenum Press, New York.

Pârvu, R., Podină, C., Zaharescu, T., and Jipa, S. 2010. Stability evaluation of polyurethane coatings by gamma irradiation. *Optoelectron. Adv. Mater.* 4(11):1815–1818.

Qi, S., Zhang, C., and Yang, X. 2004. Experimental study of photoinduced birefringence in azo-dye-doped polymer. *Optik* 115:253–256.

Reichmanis, E., Frank, C.W., O'Donnell, J.H., and Hill, D.J.T. 1993. Radiation effects on polymeric materials—A brief overview. In *Irradiation of Polymeric Materials*, eds. E. Reichmanis, C.W. Frank, and J.H. O'Donnell, 1–8. *ACS Symposium Series*; American Chemical Society, Washington, DC.

Roy, M.K., Mahloniya, R.G., Bajpai, J., and Bajpai, A.K. 2012. Spectroscopic and morphological evaluation of gamma radiation irradiated polypyrole based nanocomposites. *Adv. Mater. Lett.* 3(5):426–432.

Şakar-Deliormanli, A. 2013. Effect of cationic polyelectrolyte on the flow behavior of hydroxypropyl methyl cellulose/polyacrylic acid interpolymer complexes. *J. Macromol. Sci. Part B Phys.* 52:1531–1544.

Schmitz, J.V. 1966. *Testing of Polymers*, vol. I. John Wiley & Sons, Inc., New York.

Schnabel, W. 1981. *Polymer Degradation—Principles and Practical Applications.* Macmillan Publishing Co, New York, USA.

Şen, M., Uzun, C., Kantoglu, O., Erdogan, S.M., Deniz, V., and Guven, O. 2003. Effect of gamma irradiation conditions on the radiation-induced degradation of isobutylene–isoprene rubber. *Nucl. Instrum. Methods Phys. Res. Sect. B: Beam Interact. Mater. At.* 208:480–484.

Swallow, A.J. 1960. Radiation chemistry of organic compounds. In *International Series of Monographs on Radiation Effects in Materials*: pp. 148–169, ed. A. Charlesby, vol. 2. Permagon Press, Oxford.

Tagaya, A., Ohkita, H., Harada, T., Ishibashi, K., and Koike, Y. 2006. Zero-birefringence optical polymers. *Macromolecules* 39:3019–3023.

Tsuruta, T., Hayashi, T., Kataoka, K., and Ishihara, K. 1993. *Biomedical Applications of Polymeric Materials*, 1st edition. CRC Press, Inc., Tokyo.

Vinhas, G.M., Souto-Maior, R.M., de Almeida, Y.M.B., and Neto, B.B. 2004. Radiolytic degradation of poly (vinyl chloride) systems. *Polym. Degradation Stab.* 86(3):431–436.

Zhang, L., Wu, W., and Wang, J. 2007. Immobilization of activated sludge using improved polyvinyl alcohol (PVA) gel. *J. Environ. Sci. (China)* 19(11):1293–1297.

10 Role of UV-Vis Radiations in Analysis of Polymer Systems for Drug Delivery Applications

Ana Cazacu

CONTENTS

Abstract

The recent advances made in the pharmaceutical area allowed the improvement of polymeric-controlled drug release systems design, providing thus an enhanced therapeutic efficacy and reduced side effects of a required drug. This chapter presents the factors that must be taken into account before developing controlled release systems, a classification of these systems, and an overview of the release mechanisms. An important aspect consists in the identification and understanding of the involved mechanisms during the drug release process, since usually a combination of different mechanisms is found. Thereafter, as examples of efficient controlled drug release systems obtaining, two types of polymer-based systems are presented: chitosan matrices containing tannic acid and chitosan liposomes containing ketoprofen. The first type was studied *in vitro* by evaluating the drug release kinetics, at ~37°C, in three dissolution media having different pH values (2.0, 5.2, and 9.3). The tannic acid release mechanism from chitosan matrices was identified by using the Korsmeyer–Peppas model. The second type of controlled drug release system was studied at first *in vitro*, and then *in vivo* in order to assess the ketoprofen analgesic effect on mice nociception. For this, a hot-plate analgesia meter was involved to measure the response latency time after mice paws thermal stimulation.

10.1 INTRODUCTION

In the last few decades, a great deal of efforts to develop sustained-release drug delivery system has been made, following the need to obtain systems capable of efficient delivery of the active ingredient to the target organ. Design and synthesis of these systems are of high importance since it is aimed to obtain a longer duration of action, optimum biological response, low toxicity, and, also, to decrease the used dosage as compared to conventional dosage forms (Webster et al. 2013).

Some disadvantages that may occur and should be considered include possible toxicity or lack of compatibility of the materials used, unwanted degradation of by-products, possible surgical interventions for implantation or removal of the system, possible patient discomfort due to the chosen drug carrier, and the higher cost of controlled drug release systems compared to traditional pharmaceutical formulations.

Oral conventional dosage formulations rapidly release the drug and its absorption occurs immediately, requiring re-administration at regular intervals and, in consequence, increasing patient discomfort (Zhang et al. 2013). Furthermore, there is a risk of under- or overdosing, because for all drugs, there is a minimum concentration below which the therapeutic effect is insufficient and a maximum one above which toxic effects may be induced.

By using controlled drug release systems, it is attempted to maintain a constant certain concentration of the drug in blood or targeted tissue, over an extended period of time. The drug duration of action may be altered by modifying the biological system, the drug, or the pharmaceutical dosage form. Changing the dosage form has proved to be the most successful and, therefore, the most applied method to obtain a prolonged action of the drug (Robinson and Lee 1987).

In order to have a constant level of drug within the body, the drug release rate should be equal to the rate of inactivation and elimination. This can be achieved by modifying the drug molecule in a way that will decrease the elimination rate or by optimizing the system to obtain a slower release rate, and, thus, a decreased rate of absorption.

As regards the tailoring of controlled drug release systems, this should be done by taking into account the following aspects: the intended route for drug delivery, the targeted sites, the disease, and condition of the patient. In addition, the physicochemical properties of the drug (water solubility, molecular size, stability, partition coefficient, and protein binding) and the biological factors (absorption, distribution, metabolism, duration of action, etc.) should also be considered (Robinson and Lee 1987).

Controlled drug delivery systems are an important alternative to conventional formulations when it is necessary to slow down the water-soluble drugs release, obtain a rapid release of low-solubility drugs, deliver the drug to specific locations in the body, transport two or more active substances with similar pharmaceutical structures, and use carriers that can dissolve or degrade and be quickly eliminated (Majeti and Kumar 2000).

The most important aspect in the development of drug delivery systems is to choose appropriate polymers to be involved in the release matrix achieving. It is imperative for the next requirements to be met:

- A biocompatible polymer must be used. Also, the polymer must be able to degrade into smaller fragments inside the body, which can be excreted afterward. If not, the polymer matrix has to be surgically removed after drug releasing (Yu et al. 1997).
- The products of polymer degradation should be nontoxic and should not produce inflammatory responses.
- The polymer degradation has to take place in a reasonable period of time.

The polymer plays an essential role in protecting the active ingredient during transportation to the target and in controlling the drug release.

10.2 TYPES OF CONTROLLED DRUG DELIVERY SYSTEMS

The main focus of studies related to drug delivery applications is on self-assembled nanoparticles based on polymers, lipids, or inorganic materials, with different physicochemical properties, in which drugs may be encapsulated, adsorbed, or dispersed (Liggins and Burt 2002, Gabizon et al. 2003, Klumpp et al. 2006, Paciotti et al. 2006).

Although the diameters of nanoparticles vary between 1 and 1000 nm, not all of them are suitable as drug carriers. Thus, nanoparticles should have a diameter larger than 10 nm to avoid renal clearance (Davis 2009), but smaller than 200 nm to obtain a prolonged time of circulation within the body (Moghimi et al. 1993).

Drug delivery systems can be classified mainly into micelles, nanospheres, nanocapsules, polymersomes, and liposomes.

Micelles are spherical aggregates formed in colloidal solutions by amphiphilic molecules, having diameters ranging from 5 to 100 nm. They present a core–shell architecture, with the shell consisting of hydrophilic regions of the molecules and the core of hydrophobic parts of the molecules. Hence, the core creates a load space for the solubilization of lipophilic drugs (Lavasanifar et al. 2002).

Nanospheres can be considered as solid colloidal nanoparticles of matrix type, and are larger than micelles (diameters varying between 100 and 200 nm). The core fluidity depends on the precursor solutions used. In these nanospheres, drugs are physically and uniformly dispersed (Soppimath et al. 2001).

Nanocapsules (100–300 nm in diameter) and polymersomes (5 nm–5 μm in diameter) are vesicular systems consisting of a liquid core, in which drugs are confined, and a biopolymer membrane. *Nanocapsules* have an oily liquid as the core, and the polymer membrane is formed by a monolayer. They are used to encapsulate and deliver hydrophobic drugs (Teixeira et al. 2005). *Polymersomes* have an aqueous phase as the core, a bilayer polymer as the membrane, and are used to entrap and deliver water-soluble drugs (Discher and Eisenberg 2002).

Liposomes (50–200 nm in diameter) are analogous to polymersomes, but the difference is that liposomes consist of one or several concentric bilayers of lipid (biological amphiphilic molecules with two hydrophobic tails). Liposomes with a single bilayer (or lamella) are named unilamellar vesicles (ULV), and the ones with several bilayers are named multilamellar vesicles (MLV) (Evans and Wennerstrom 2001).

In preparing liposomes, it is essential to choose the right concentration of lipid to be dissolved in an organic solvent. The bilayers will self-assemble after adding aqueous solutions (due to the hydrophobic tails tendency to minimize contact with water at the bilayer ends), only if the concentration is low. At high concentrations, bilayers will transform into a lamellar phase.

Another important class of controlled drug delivery system is represented by *hydrophilic polymer matrix systems* with constant release rate, which are frequently used for oral intake (Colombo et al. 2000, Prajapati and Patel 2010). Hydrophilic polymers swell when they are exposed to an aqueous medium and form a gel layer (acting like a barrier) on the surface, from which the drug is released by diffusion and erosion of the polymer (Maderuelo et al. 2011).

10.3   MECHANISMS OF CONTROLLED RELEASE

Drug release from delivery systems is predominantly regulated by one or a combination of the subsequent mechanism: diffusion, swelling, erosion, and degradation.

Diffusion is characteristic to nondisintegrating matrix devices and implies the active substance passing through the polymeric wall of the controlled-release system into the external medium. This mechanism can take place on a macroscopic scale by drug diffusion through matrix pores or on a molecular level by passing between polymer chains.

Swelling occurs when the polymer systems uptake water or other body fluids, leading to an increased volume, and, as a result, to the activation of the confined drugs. The drug release rate can be limited by diffusion (if the swelling is rapid) or by swelling itself (if it is slow enough).

Swelling-controlled release systems are based on hydrogels that are capable of swelling without dissolving when placed in aqueous solvents, their absorption capacity of fluid being up to 60%–90% (Kim and Chu 2000). Depending on the nature of the selected polymer, swelling can be triggered by changes in the surrounding medium, such as pH, temperature, or ionic strength.

In order to optimize drug delivery systems, there is a high interest on *erodible* and *degradable* biopolymers. This kind of polymers do not need to be removed from the body after completely releasing the active agent, because they will be degraded by the natural biological processes into nontoxic and excretable or resorbable smaller compounds.

Erodible systems may be subject to two major mechanisms: bulk erosion and surface erosion. Bulk erosion occurs when water diffuses into a polymer matrix with a higher rate than the rate of hydrolysis. Consequently, the whole material starts to degrade due to the hydrolysis process taking place from inside out, and mass loss is noticed. Surface erosion is characterized by the degradation of a polymer matrix at the exterior surface, and not on the inside of the material. In this case, the rate of polymer degradation is higher than the rate of water diffusion into the matrix, leading to a system degraded only on the surface (Ulery et al. 2011).

10.4 *IN VITRO* STUDY OF TANNIC ACID RELEASE KINETICS FROM POLYMER MATRICES

Among the controlled delivery dosage formulations, polymer matrix systems that incorporate active substances attract much interest due to the ease of manufacturing and designing to obtain the desired release profile.

Therefore, this study is focused on matrices obtained from chitosan (a natural polymer) by dry-phase inversion method, which incorporate tannic acid as a polyphenol model, and in the presence of a cationic surfactant. The aim is to achieve a drug-containing matrix that presents zero-order release kinetics. In order to analyze the drug entrapment efficiency of the corresponding polymer matrices and the release kinetics, the simplest and faster way is to use ultraviolet-visible (UV-Vis) spectrometry, which is a reliable method widely employed.

In a previous study (Garlea et al. 2009), it was shown that chitosan can be efficiently used in combination with cetyltrimethylammonium bromide (CTAB) as a cationic surfactant to form matrices for entrapping drugs. Both chitosan and CTAB being cationic compounds, it was concluded that the self-assembly of the polymer matrix is strongly dependent on drug presence.

Polyphenols are antioxidants, and they find use in combating oxidative stress, which can lead to neurodegenerative and cardiovascular diseases, osteoporosis, or diabetes (Graf et al. 2005). Polyphenols are found in many legumes, fruits, beverages, and cereals.

10.4.1 MATERIALS AND METHODS

The following materials were used: chitosan (79.7% *N*-deacetylation degree and a molecular weight of 310,000 g/mol) from Vanson Chemicals, Redmond, WA, USA; tannic acid (molecular weight of 1701.20 g/mol) from Sigma Aldrich; CTAB from Chemapol; acetic acid (99.5%) from Chemical Company; and Milli-Q water (18.2 M$\Omega \cdot$ cm).

The preparation protocol of 2% and 3% chitosan solutions with 6 mm CTAB and different tannic acid concentrations (5, 7, and 10 mm) is given elsewhere (Garlea et al. 2010).

The release kinetics of tannic acid from these polymeric matrices was studied with a NanoDrop-1000 spectrophotometer that has the advantage of using very small volumes from a sample, that is, of only 1 μL. For this reason, both the eluent concentration and the process are considered unaltered.

In order to evaluate the release kinetics, three dissolution media with different pH values were used: hydrochloric acid (pH = 2.0), phosphate buffer solution (pH = 5.2), and urea (pH = 9.3). Thus, 20 mg from each matrix were placed into 50 mL of different dissolution medium, and the drug absorbance was assessed at predetermined time intervals. The release kinetics was performed for 32 h at 37 $\pm$ 0.5°C. All the measurements were done at a wavelength of 277 nm (the maximum band of the tannic acid absorption spectrum). At this value, the polymer or the dissolution medium shows no absorption of light.

10.4.2   RESULTS AND DISCUSSIONS

For further calculations of the drug concentration released from the polymeric matrix, the tannic acid calibration curve was obtained (Figure 10.1).

The data depicted in Figure 10.1 are characterized by the following equation:

$$A = 0.043 \times C \tag{10.1}$$

where A represents the absorbance and C is the concentration of tannic acid.

Also, it was found that the correlation coefficient describing the linear equation is $r = 0.999$, indicating that the regression line fits the data very well and the degree of accuracy is very good.

The released concentration of tannic acid from the CTAB–chitosan matrices was assessed by measuring the absorbance of the dissolution medium at 277 nm. Furthermore, it is important to note that the maximum wavelength of the tannic acid was the same, regardless of the release medium used. This fact proves that the presence of the polymer matrix did not interfere with the drug absorbance at this wavelength.

Over the years, several theoretical models have been given to explain the processes involved in water or drug transportation through a polymer matrix, and to predict the appropriate release kinetics of that drug (Korsmeyer et al. 1986a,b, Cohen and Erneux 1988a,b, Ju et al. 1995, Siepmann et al. 1999a,b). A precise mathematical depiction of the release kinetics is a complex task since there are numerous parameters that must be considered, such as matrix dimension and porosity, the water

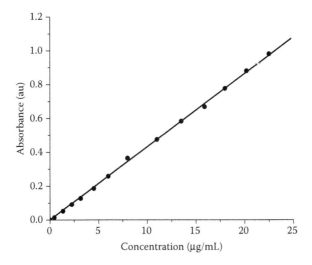

FIGURE 10.1   Tannic acid calibration curve ($\lambda = 277$ nm).

diffusion into the polymer matrix and drug diffusion out of the matrix, polymer dissolution and swelling, species diffusivity depending on concentration, etc.

For tannic acid release mechanism from chitosan matrices, the simple relationship given by Korsmeyer was used (Korsmeyer et al. 1983)

$$\frac{M_t}{M_\infty} = kt^n \qquad (10.2)$$

where M_t is the amount of drug released at time t, M_∞ is the amount of drug released
at infinite time, t is the release time, k is the rate constant characteristic to the drug–
polymer system, and n is the release exponent. The value of n from Equation 10.2
characterizes the type of the release mechanism. Hence, the release mechanism is
described by Fickian diffusion if $n = 0.45$, by anomalous (non-Fickian) diffusion if
$0.45 < n < 0.89$, and by zero-order release if $n = 0.89$ (the drug release rate is not
concentration dependent).

To identify the tannic acid release mechanism from chitosan matrices, Korsmeyer–
Peppas model was used to fit the drug release data comprised in the interval of
10%–70%.

These types of polymer matrices undergo a three-step process to release the drug
contained inside: hydration, swelling, and erosion. The drug is then transported from
the hydrated matrix or from the eroded parts to the corresponding medium of dissolution (Kiortsis et al. 2005).

In Figures 10.2 through 10.4, the kinetic curves for tannic acid release from different matrices based on chitosan are shown.

The kinetic curves indicate that the release of tannic acid from chitosan matrices
differs from one dissolution medium to another. In hydrochloric acid (pH = 2), all
the matrices degraded after 32 h, while in phosphate buffer solution (pH = 5.2) and
urea (pH = 9.3), they were only partially degraded during that time. The results show
a complex mechanism of erosion and diffusion, which is, obviously, faster in the
hydrochloric acid medium.

The release kinetic parameters of the tannic acid calculated by using the
Korsmeyer–Peppas mathematical model are given in Table 10.1.

The data obtained using the Korsmeyer–Peppas model indicate a non-Fickian
diffusion release mechanism (anomalous diffusion), as well as a zero-order kinetics
in case of tannic acid release in an acidic medium.

Based on the calibration curve, the percent content of released drug from the 2%
chitosan_6 mm CTAB_10 mm tannic acid sample was of 90% (pH 2), 20% (pH 5.2),
and 25% (pH 9.3). Similarly, for the 3% chitosan_6 mm CTAB_10 mm tannic acid
sample, the following values were obtained: 90% (pH 2), 25% (pH 5.2), and 31% (pH
9.3). The best results were found in the acidic medium, because the erosion process
(which is rate restrictive) is enhanced in this extreme pH value.

10.4.3 Conclusions

The release kinetic analysis performed on CTAB–chitosan matrices containing
tannic acid as the active substance demonstrates that they are good candidates

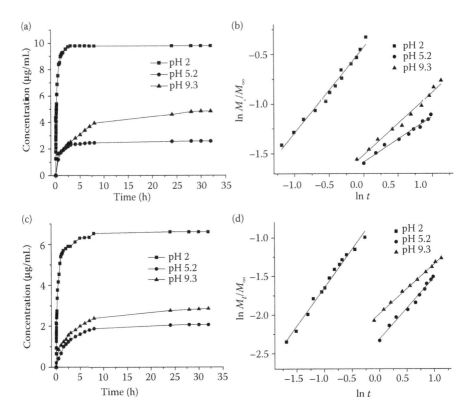

FIGURE 10.2   Kinetic curves and corresponding fitted data for tannic acid release from the matrices of 2% (a, b) and 3% (c, d) chitosan with 6 mm CTAB and 5 mm tannic acid.

for controlled drug release formulations, an appropriate dissolution profile being achieved.

Furthermore, since these matrices are pH-sensitive, their design can be optimized so as to present zero-order release kinetics both in the stomach and intestine for a much longer period of time than in the conventional dosage formulation case.

10.5   *IN VIVO* STUDY OF THE ANALGESIC EFFECT OF KETOPROFEN POLYMER LIPOSOMES

In order to confirm the correct choice of using a certain type of a controlled drug delivery system, besides the stability analysis of the system and the *in vitro* investigation, an *in vivo* study must be made. For this reason, hereinafter, a study aiming the physicochemical properties investigation of ketoprofen soft-matter liposomes coated with chitosan, as well as their *in vivo* effects analysis on mice nociception are presented.

Ketoprofen is a nonsteroidal anti-inflammatory (NSAIDs) drug prescribed to relieve the pain, fever, and inflammation (presents analgesic, antipyretic, and anti-inflammatory effects). Designing liposomes as carriers for NSAIDs is of great

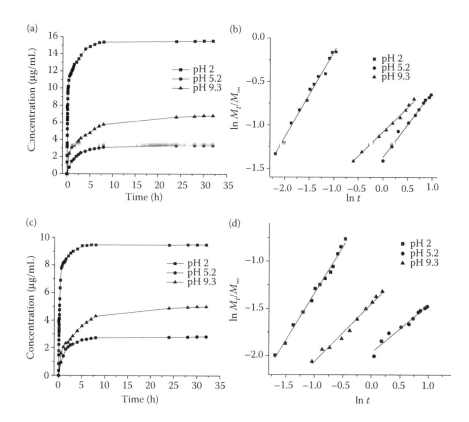

FIGURE 10.3 Kinetic curves and corresponding fitted data for tannic acid release from the matrices of 2% (a, b) and 3% (c, d) chitosan with 6 mm CTAB and 7 mm tannic acid.

interest since this type of drugs cause gastrointestinal or renal side effects, and can generate esophageal, stomachal, and intestinal lesions (McCarthy 2000).

Chitosan, a natural biodegradable polymer, was selected due to the results obtained in a previous study (Garlea et al. 2007) that showed the role played by this as a stabilizer and additional barrier for the self-assembled liposomes containing ketoprofen.

10.5.1 Materials and Methods

As materials were used: L-α-phosphatidylcholine (Egg Yolk PC 99% TLC purity), ketoprofen (98% purity, Mw = 254.28 g/mol), ethyl alcohol and acetic acid from Sigma-Aldrich, chitosan (79.7% N-deacetylation degree, Mw = 310,000 g/mol) from Vanson Chemicals, Redmond, WA, USA, and Milli-Q water (18.2 MΩ·cm).

A detailed description of the preparation protocol of chitosan-coated liposomes with ketoprofen is given elsewhere (Tartau et al. 2012). The obtained liposomes were subjected to dialysis for 10 h until a physiological pH of about 7 was reached, by using a Sigma D6191-25EA dialysis-tubing cellulose membrane. To characterize the electrochemical equilibrium on interfaces of all the liposomes, a Malvern Zetasizer

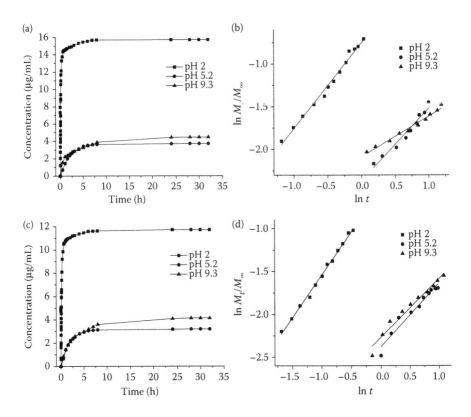

FIGURE 10.4   Kinetic curves and corresponding fitted data for tannic acid release from the matrices of 2% (a, b) and 3% (c, d) chitosan with 6 mm CTAB and 10 mm tannic acid.

Nano ZS, ZEN-3500 model was employed. By this method, the surface charge of liposomes was assessed from the zeta potential (ZP) data that can be considered as a good indicator for the colloidal particles interactions and their stability (Müller et al. 2001).

The *in vitro* drug release analysis was performed on a Hewlett Packard 8453 UV-Vis spectrophotometer, in the range of 190–500 nm. The absorbance was measured at 258 nm (the maximum absorption wavelength of ketoprofen); a calibration curve of absorbance versus ketoprofen concentration was plotted.

The *in vivo* tests were done on white Swiss male mice (20–25 g), divided into three groups of seven animals each. The mice were treated using an eso-gastric device as follows: the first group (control) received 0.3 mL of Milli-Q water, the second group (ketoprofen) was treated with 10 mg/kg of body weight of ketoprofen, and the third group (ketoprofen vesicles) with 10 mg/kg of body weight of dialyzed chitosan–ketoprofen liposomes.

For the nociceptive cutaneous stimuli evaluation, a behavioral acute pain model was tested by utilizing an Ugo Basile hot-plate analgesia meter to measure the response latency time after mice paws thermal stimulation. This was done at specific moments, namely 15, 30, 60, and 90 min; 2, 4, 6, 8, 10, and 12 h after oral administration.

TABLE 10.1
Drug Release Kinetic Parameters from Chitosan Matrices

Sample	pH	n	$k\ (h^{-n})$
2% chitosan_6 mm CTAB_5 mm tannic acid	2	0.86	0.37
	5.2	0.46	0.03
	9.3	0.63	0.03
3% chitosan_6 mm CTAB_5 mm tannic acid	2	0.99	0.22
	5.2	0.92	0.01
	9.3	0.64	0.01
2% chitosan_6 mm CTAB_7 mm tannic acid	2	0.98	6.75
	5.2	0.75	0.04
	9.3	0.56	0.08
3% chitosan_6 mm CTAB_7 mm tannic acid	2	0.99	0.46
	5.2	0.51	0.01
	9.3	0.61	0.04
2% chitosan_6 mm CTAB_10 mm tannic acid	2	1	0.18
	5.2	0.84	0.01
	9.3	0.47	0.01
3% chitosan_6 mm CTAB_10 mm tannic acid	2	1	0.29
	5.2	0.75	0.01
	9.3	0.69	0.01

The statistical analysis of obtained data was performed with SPSS software for Windows and ANOVA method. In this experiment, the p-values less than 0.05 as compared to the ones of the control group are considered statistically significant.

Experiments were carried out in agreement with the recommendations and policies of the EU Directive 2010/63/EU.

10.5.2 RESULTS AND DISCUSSIONS

In Table 10.2, the average diameters and the average ZP of the samples are presented. As can be noticed, the ketoprofen molecules dispersed in aqueous solution have an average diameter of 183 nm, while the liposomes loaded with ketoprofen show a slight increase, their average diameter being of 222 nm. By chitosan addition to the

TABLE 10.2
Average Values of Samples Diameter and Zeta Potential

System	Average Diameter (nm)	Average ZP (mV)
Ketoprofen	183	−0.177
Ketoprofen liposomes	222	−23.4
Chitosan–ketoprofen liposomes	1850	+9.88
Dialyzed chitosan–ketoprofen liposomes	1287	+20.3

self-assembled ketoprofen liposomes solution, a significant enlargement of the average diameter (1850 nm) of the achieved vesicles was observed. In case of dialyzed chitosan–ketoprofen liposomes, a decrease in diameter was recorded as a result of the aging process (Tartau et al. 2012).

ZP is a predictive measure of long-term stability of colloidal dispersions. If the particles from the colloidal suspension possess a high value (positive or negative) of ZP, then, the particles will not tend to aggregate due to the electric or steric repulsive forces between each other, and the system will be stable. If the value will be low, the attractive forces between particles will be predominant and the system will flocculate.

A ZP value of at least ±30 mV for electrostatically or ±20 mV for sterically stabilized particles indicates a stable suspension (Müller et al. 2001). Therefore, from Table 10.2, it can be observed that both ketoprofen liposomes (ZP = –23.4 mV) and dialyzed chitosan–ketoprofen liposomes (ZP = +20.3 mV) suspensions are stable. Chitosan induced a positively charging after its adsorption on the liposomes surface. Polymer layers on particles act like a barrier against the van der Waals attraction, which leads to steric repulsion.

The UV-Vis spectrum from Figure 10.5, showing a diminished absorption band for drug-containing liposomes (at 258 nm), indicates an encapsulation degree of ketoprofen into the chitosan vesicles of 95%, as compared to the free ketoprofen.

In order to depict the variation in time of ketoprofen molar concentration, the release rate (R) was defined as the number of moles of ketoprofen released per unit time and per unit volume of release medium

$$R(t) = \frac{1}{V} d\nu(t) \qquad (10.3)$$

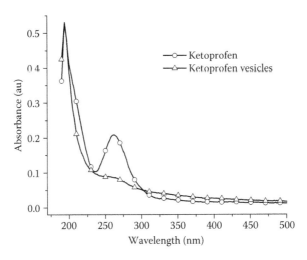

FIGURE 10.5 UV-Vis spectra for free ketoprofen and chitosan-coated liposomes with ketoprofen.

where V is the volume, dv is the number of moles of ketoprofen released during dt. The molar concentration at a certain point in time (t) was expressed by the relation

$$C(t) = \frac{v(t)}{V} \qquad (10.4)$$

where C is the molar concentration of ketoprofen in the release environment, and v the number of moles of ketoprofen existing at t time in the V volume.

The ketoprofen *in vitro* release profile (Figure 10.6) showed that after 1 h, 21% of ketoprofen was released from lipid vesicles in the medium. Approximately 50% of the entrapped ketoprofen was released after 3 h, and 98% after 38 h. This profile demonstrates a prolonged release of ketoprofen from vesicles.

From the statistical analysis of the recorded data during the *in vivo* antinociceptive evaluation on mice (Figure 10.7), it can be seen that oral administration of ketoprofen (ketoprofen group) resulted in a rapid increase of the response latency time, shortly after paw thermal noxious stimulation. The most evident effect was noticed between 30 and 90 min, but continued until 120 min after drug administration. The response latency time was prolonged to 180 after thermal stimulation, but no significant results were obtained in comparison with those of the control group.

By administering a solution of dialyzed chitosan–ketoprofen liposomes (ketoprofen vesicles group), the response latency time was significantly increased after 120 min from thermal paws stimulation. The analgesic effect was extended for more than 10 h, with a peak in the interval between 2 and 8 h, as compared to the control group. The maximum possible effect of dialyzed chitosan–ketoprofen liposomes was comparable with that of nonentrapped ketoprofen, but the time interval was significantly longer.

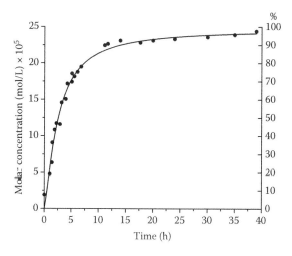

FIGURE 10.6 *In vitro* release profile of ketoprofen from vesicles.

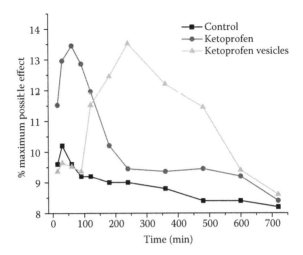

FIGURE 10.7 The response latency time of the ketoprofen and dialyzed chitosan–ketoprofen liposomes (ketoprofen vesicles) during the hot-plate test. Each point represents the mean ± standard error of mean of latency time for seven mice. $*p < 0.05$ versus control.

10.5.3 Conclusions

It was shown that chitosan is able to stabilize the obtained liposomes, and to change the size of them and the ZP of the solution. The liposomes' size was increased after the adsorption of an additional chitosan layer.

The *in vivo* hot-plate tests evidenced a prolonged antinociceptive effect in case of dialyzed chitosan–ketoprofen liposomes, having approximately the same intensity as compared to the nonentrapped substance, but the release was sustained for a longer period of time.

By this method, a new chitosan-coated–lipid dosage formulation entrapping ketoprofen was developed, which is suitable for its use as controlled-release drug delivery system.

10.6 CONCLUDING REMARKS

In this chapter, a variety of controlled drug release systems and the fundamental mechanisms of release were presented. Furthermore, two different types of such systems were explained in detail as models of inexpensive controlled-release products. For their investigation, both *in vitro* and *in vivo* approaches were taken into account.

The presented data show the importance of UV-Vis radiations in analyzing drug delivery systems, the spectrophotometric method being a facile, rapid, and sensitive technique to be employed for this purpose.

REFERENCES

Cohen, D.S. and Erneux, T. 1988a. Free boundary problems in controlled release pharmaceuticals. I: Diffusion in glassy polymers. *SIAM J. Appl. Math.* 48:1451–1465.

Cohen, D.S. and Erneux, T. 1988b. Free boundary problems in controlled release pharmaceuticals. II: Swelling controlled release. *SIAM J. Appl. Math.* 48:1466–1474.

Colombo, P., Bettini, R., Santi, P., and Peppas, N.A. 2000. Swellable matrices for controlled drug delivery: Gel-layer behaviour, mechanisms and optimal performance. *Pharm. Sci. Technol. Today* 3:198–204.

Davis, M.E. 2009. The first targeted delivery of siRNA in humans via a selfassembling, cyclodextrin polymer-based nanoparticle: From concept to clinic. *Mol. Pharm.* 6:659–668.

Discher, D.E. and Eisenberg, A. 2002. Polymer vesicles: Materials science: Soft surfaces. *Science* 297:967–973.

Evans, D.F. and Wennerstrom, H. 2001. *The Colloidal Domain: Where Physics, Chemistry, Biology, and Technology Meet.* Wiley-VCH, New York.

Gabizon, A., Shmeeda, H., and Barenholz, Y. 2003. Pharmacokinetics of pegylated liposomal doxorubicin: Review of animal and human studies. *Clin. Pharmacokin.* 42:419–436.

Garlea, A., Manole, A., Popa, M.I., and Melnig, V. 2009. Chitosan–paracetamol nanostructure self-assembling matrices as drug delivery systems. *Mat. Plast.* 46(4):356–362.

Garlea, A., Melnig, V., and Popa, M.I. 2010. Nanostructured chitosan–surfactant matrices as polyphenols nanocapsules template with zero order release kinetics. *J. Mater. Sci. Mater. Med.* 21(4):1211–1223.

Garlea, A., Popa, M.I., Pohoata, V., and Melnig, V. 2007. Ibuprofen/ketoprofen entrapment in chitosan based vesicle carrier. *Rom. J. Biophys.* 17(3):157–168.

Graf, B.A. Milbury, P.E., and Blumberg, J.B. 2005. Flavonols, flavonones, flavanones and human health: Epidemiological evidence. *J. Med. Food* 8:281–290.

Ju, R.T.C., Nixon, P.R., Patel, M.V., and Tong, D.M. 1995. Drug release from hydrophilic matrices. 2. A mathematical model based on the polymer disentanglement concentration and the diffusion layer. *J. Pharm. Sci.* 84:1464–1477.

Kim, S.H. and Chu, C.C. 2000. Pore structure analysis of swollen dextran–methacrylate hydrogels by SEM and mercury intrusion porosimetry. *J. Biomed. Mater. Res.* 53:258–266.

Kiortsis, S., Kachrimanis, K., Broussali, Th., and Malamataris, S. 2005. Drug release from tableted wet granulations comprising cellulosic (HPMC or HPC) and hydrophobic component. *Eur. J. Pharm. Biopharm.* 59:73–83.

Klumpp, C., Kostarelos, K., Prato, M., and Bianco, A. 2006. Functionalized carbon nanotubes as emerging nanovectors for the delivery of therapeutics. *Biochim. Biophys. Acta* 1758:404–412.

Korsmeyer, R.W., Gurny, R., Doelker, E., Buri, P., and Peppas, N.A. 1983. Mechanisms of solute release from porous hydrophilic polymers. *Int. J. Pharm.* 15:25–35.

Korsmeyer, R.W., Lustig, S.R., and Peppas, N.A. 1986a. Solute and penetrant diffusion in swellable polymers. I. Mathematical modeling. *J. Polym. Sci. Polym. Phys.* 24:395–408.

Korsmeyer, R.W., von Meerwall, E., and Peppas, N.A. 1986b. Solute and penetrant diffusion in swellable polymers. II. Verification of theoretical models. *J. Polym. Sci. Polym. Phys.* 24:409–434.

Lavasanifar, A., Samuel, J., and Kwon, G.S. 2002. Poly(ethylene oxide)-blockpoly(L-amino acid) micelles for drug deliver. *Adv. Drug Deliv. Rev.* 54:169–190.

Liggins, R.T. and Burt, H.M. 2002. Polyether–polyester diblock copolymers for the preparation of paclitaxel loaded polymeric micelle formulations. *Adv. Drug Deliv. Rev.* 54:191–202.

Maderuelo, C., Zarzuelo, A., and Lanao, J.M. 2011. Critical factors in the release of drugs from sustained release hydrophilic matrices. *J. Control. Release* 154:2–19.

Majeti, N.V. and Kumar, R. 2000. Nano and microparticles as controlled drug delivery devices. *J. Pharm. Pharmaceut. Sci.* 3(2):234–258.

McCarthy, M. 2000. Comparative toxicity of nonsteroidal anti-inflammatory drugs. *Am. J. Med.* 107:37–46.

Moghimi, S.M., Hedeman, H., Muir, I.S., Illum, L., and Davis, S.S. 1993. An investigation of the filtration capacity and the fate of large filtered sterically-stabilized microspheres in rat spleen. *Biochim. Biophys. Acta* 1157:233–240.

Müller, R.H., Jacobs, C., and Kayser, O. 2001. Nanosuspensions as particulate drug formulations in therapy. Rationale for development and what we can expect for the future. *Adv. Drug Deliv. Rev.* 47:3–19.

Paciotti, G.F., Kingston, D.G.I., and Tamarkin, L. 2006. Colloidal gold nanoparticles: A novel nanoparticle platform for developing multifunctional tumor-targeted drug delivery vectors. *Drug Dev. Res.* 67:47–54.

Prajapati, G.B. and Patel, R.K. 2010. Design and *in vitro* evaluation of novel nicorandil sustained release matrix tablets based on combination of hydrophilic and hydrophobic matrix systems. *Int. J. Pharm. Sci. Rev. Res.* 1:33–35.

Robinson, J.R. and Lee, V.H.L. 1987. *Controlled Drug Delivery: Fundamentals and Applications*. Marcel Dekker, New York.

Siepmann, J., Kranz, H., Bodmeier, R., and Peppas, N.A. 1999a. HPMC-matrices for controlled drug delivery: A new model combining diffusion, swelling and dissolution mechanisms and predicting the release kinetics. *Pharm. Res.* 16:1748–1756.

Siepmann, J., Podual, K., Sriwongjanya, M., Peppas, N.A., and Bodmeier, R. 1999b. A new model describing the swelling and drug release kinetics from hydroxypropyl methylcellulose tablets. *J. Pharm. Sci.* 88:65–72.

Soppimath, K.S., Aminabhavi, T.M., Kulkarni, A.R., and Rudzinski, W.E. 2001. Biodegradable polymeric nanoparticles as drug delivery devices. *J. Control. Release* 70:1–20.

Tartau, L., Cazacu, A., and Melnig, V. 2012. Ketoprofen–liposomes formulation for clinical therapy. *J. Mater. Sci. Mater. Med.* 23(10):2499–2507.

Teixeira, M., Alonso, M.J., Pinto, M.M.M., and Barbosa, C.M. 2005. Development and characterization of PLGA nanospheres and nanocapsules containing xanthone and 3-methoxyxanthone. *Eur. J. Pharm. Biopharm.* 59:491–500.

Ulery, B.D., Nair, L.S., and Laurencin, C.T. 2011. Biomedical applications of biodegradable polymers. *J. Polym. Sci. B Polym. Phys.* 49:832–864.

Webster, D.M., Sundaram, P., and Byrne, M.E. 2013. Injectable nanomaterials for drug delivery: Carriers, targeting moieties, and therapeutics. *Eur. J. Pharm. Biopharm.* 84:1–20.

Yu, J.L., Johansson, S., and Ljungh, A. 1997. Fibronectin exposes different domains after adsorption to a heparinized and an unheparinized poly(vinyl chloride) surface. *Biomaterials* 18:421–427.

Zhang, Y., Chan, H.F., and Leong, K.W. 2013. Advanced materials and processing for drug delivery: The past and the future. *Adv. Drug Deliv. Rev.* 65:104–120.

11 Interaction of Radiations with Stretched Polymer Foils in Controlling the Release of a Drug for Alzheimer's Disease

Beatrice Carmen Zelinschi

CONTENTS

11.1 INTRODUCTION

Over the last two decades, significant progress has been made in the development of biocompatible and biodegradable materials used in biomedical applications. For the biomedical field, the goal is to develop and characterize artificial materials or, in other words, "spare parts" for use in the human body to measure, restore, and improve physiologic functions, and enhance survival and quality of life. Typically, inorganic and polymeric materials have been used as artificial heart valves, synthetic blood vessels, artificial hips, medical adhesives, sutures, dental composites, and polymers for controlled slow drug delivery. The development of new biocompatible materials includes considerations that go beyond nontoxicity to bioactivity as it relates to interacting with and, in time, being integrated into the biological environment as well as other tailored properties depending on the specific "*in vivo*" application.

The controlled-release systems allow the drug/active substance to be released in a reproducible, predictable, and predetermined way. In the systems with controlled delivery, the active substance is dispersed in a polymer matrix from which it is released after a kinetic profile corresponding to a given treatment of the disease. To achieve a control-released system, in which an active substance and a polymeric matrix deliver a product that is the most compatible with the human body, we must conduct a suitable kinetic profile. A drug administrated traditionally to a patient will release the active substance immediately, after incorporation into the body, and the diminution of the concentration takes place; so, a new dose of drug must be administrated periodically (Steele and Glazier 1999). The controlled release of a therapeutical agent facilitates the maintenance of a constant and optimal dose during the treatment. The drugs incorporated in different polymer matrices and directed to a specified target have low toxicity and eliminate the discontinuity of treatment.

Generally, the drug introduced into the body has to have a minimum level for producing the anticipated effect and a maximum level over which it becomes toxic. In the system with controlled delivery, the active substance is dispersed in a polymer matrix from which it is released after a kinetic profile corresponding to a given treatment of the disease (Rojas-Fernandez 2001, Sugimoto et al. 2002, Speight and Lange 2005). The poly (vinyl alcohol) (PVA) membranes are often used in the preparation of the different biomedical materials (dialysis membranes, cardiovascular devices, and active substance controlled-release systems). The PVA is a polymer used in the pharmaceutical technology due to the high tolerability and reduced toxicity for the patients.

The progresses in the controlled release of the drugs offer a significant degree of freedom in choosing of the target; the systems containing the drug can be placed in a cavity of the organism, or they can be attached on the skin-patch form (Schroeder et al. 2007).

11.2   BASIC MATERIALS AND METHODS USED IN CONTROLLING THE RELEASE OF A DRUG FOR ALZHEIMER'S DISEASE

Being easily manufactured and mechanically stable, PVA has a preferred role in obtaining different biomedical materials, such as membranes for dialysis, cardiovascular devices, or systems for controlled delivery. This polymer has a simple structure and it is characterized by biodegradability, biocompatibility, and water solubility. The viscosity of PVA solutions in water varies with concentration and temperature. PVA is used for biomedical applications, particularly in drug delivery, tissue replacement for improvement of human organs functionality, immunological kits, and for cancer therapy. Owing to its transparency, ease of processing, and low costs, PVA is an optical polymer widely used as a material for various optical devices. Although the PVA has a simple structure, it is biodegradable, biocompatible, and soluble in water. The ability of the substances to have many crystalline forms is called polymorphism. Polymorphism is the phenomenon by which in different physicochemical conditions (temperature, pressure, and chemical medium), various natural and artificial substances may crystallize with different internal structures, resulting in distinct crystalline species and forms belonging to different classes of symmetry. The passage of a substance from one polymorphic form to another can occur in a short time and can be reversible or irreversible.

Alzheimer's disease (AD) is a condition that causes abnormal changes in the brain mainly affecting memory and other mental abilities. Alzheimer's is a disease, which is not a normal part of aging. Loss of memory is the usual first symptom. As the disease progresses, the loss of reasoning ability, language, decision-making ability, judgment, and other critical skills make navigating day-to-day living impossible without help from others, most often a family member or friend. Sometimes, but not always, difficult changes in personality and behavior occur. No one fully understands what causes AD yet, and there is currently no cure. Considerable progress has been made by researchers in recent years though, including the development of several medications for early-stage AD which can help to improve cognitive functioning for a while. Among the main drugs, donepezil has proved to be beneficial in improving memory, with limited side effects.

Donepezil is a drug which belongs to the acetyl cholinesterase enzyme inhibitive category and it is used for the treatment of the easy and moderate forms of Alzheimer. The molecular chemical formula for donepezil is $C_{24}H_{29}NO_3$ and it is used to specifically facilitate learning or memory, particularly to prevent the cognitive deficits associated with dementias. The PVA represents the polymeric matrix where donepezil is included.

On estimate, donepezil is kinetically control released from the PVA membrane in view of acquiring information about the transdermal transfer mechanism of this drug (Rogojanu et al. 2011).

A totally hydrolyzed PVA acquired from the Merck Co., Germany was used in order to obtain the polymeric membranes. The PVA films which contain donepezil were obtained by using a PVA solution totally hydrolyzed:

- Hydrolysis degree 98%
- Molecular mass 22,000
- Viscosity measured at the temperature of 20°C comprised between 3.8 and 5.2 mPa s

Donepezil used as an active substance was acquired from India. Pure PVA films and PVA with donepezil foils have been obtained by dissolving the polymer in distilled water at the temperature of 75–80°C. The active substance— BPMDMDHI— (2-[(1-benzyl-4-piperidyl) methyl]-5,6-dimethoxy-2,3 dihydroinden-1-one) was dissolved in ethanol/acetonitrile (according to the polymorph to be obtained as a result of the crystallization) and afterward included in the PVA solution under stirring up during 1 h. The films obtained by a sediment of gel on glass supports have been let to dry during 48 h and after that kept at an exicator in ambient temperature. Water filtered by Millipak-Gradient A10, with an internal carbon filter (Quantum EX) and external Teflon filter Millipak40 of 0.22 μm was used for the preparation of the tampon solutions. The pH measurements of the tampon solutions have been done with the Mettler-Toledo pH-meter. The analysis of the particle size for the active substance (donepezil) was implemented with Malvern-Mastesizer2000 equipment, HydroS-small volume. The dispersant material used is dioctyl sulfosuccinate sodium salt (DSSS)—6.5% in isooctane. *In vitro* release studies have been carried out at the skin pH (5.5) being a matter of transdermal-controlled release. The release tests were obtained by the immersion of some PVA samples with an active substance (100 mg) into 50 mL tampon solution. At a certain period of time comprised between 0 and 24 h, a sample of 5 mL tampon solution was extracted with the drug and its ultraviolet-visible (UV-VIS) spectrum was recorded with a Varian Cary100 Bio spectrophotometer. The semicrystalline feature of PVA can be emphasized using the x-ray diffraction (XRD) analysis by realizing the diffractogram of pure PVA and then the diffractogram for donepezil subsequently embedded in the polymer matrix of PVA.

Differential scanning calorimetry (DSC), which measures the thermal capacity (or heat flux) as a function of temperature, detects and monitors thermally induced conformational transitions and phase transitions. During the analysis, depending on the sample complexity, it may appear in one or more maximum points of inflection that reflects the heat-induced transition. DSC curves identify if a process is exothermic or endothermic and also defines the types of transitions involved. If several signals appear on the chart, endothermic or exothermic, they relate to the endothermic inflection point that represents the melting substance and it is known as the inflection point. If the signals overlap, further experiments must be carried out under conditions which are different mass of the sample and the heating rate, to have a good resolution of the signals.

Thermal gravimetric (TGA) has developed itself from the classical stepwise heating and weighing a solid sample. A substance heated at different temperatures suffered a series of changes, some of which are tied to the changes in mass. Watching the mass variation as a function of temperature, one can make assumptions on the transformation of the researched material. The thermo-gravimetric analysis is a widely used method in the study of the thermal degradation of the polymers, and allows the determination of the kinetic parameters of the thermal decomposition process. The parameters may be of particular importance in elucidating mechanisms of degradation reactions and to estimate the thermal stability of polymers (Adin et al. 2006).

11.3 MATHEMATICAL PATTERNS OF THE CONTROLLED-RELEASE KINETICS

To analyze the mechanism of drug release-rate kinetics, the results of *in vitro* release profiles can be plotted in various kinetic models, such as zero order, first order, Higuchi, and Korsmeyer–Peppas. Mathematical modeling increases the understanding of the release mechanism and in turn helps to reduce the number of experiments required to optimize the formulation. The selection criteria for the most adequate model are based on the correlation coefficient. The models used for the kinetic release estimation are those taken over the literature: the kinetic model of zero order, model of order I, model Higuchi, model Korsmeyer–Peppas, and model Hixson–Crowell (Park et al. 2007, Lekshmi et al. 2012). The mathematical models are described as follows:

Zero-order release kinetics model: Describe systems where the drug release rate is constant over a period of time. The equation for zero-order release is shown in relation (11.1)

$$Q_t = Q_0 + k_0 \cdot t \tag{11.1}$$

where Q_t is the cumulative amount of drug released at time t, Q_0 is the initial amount of drug, k_0 is the release kinetic constant, and t is the time at which the drug release is calculated.

First-order model: In typical first-order release kinetics, the drug release rate depends on its concentration. The first-order release equation can be explained as it is shown in Equation 11.2

$$\ln Q_t = \ln Q_0 - k_1 \cdot t \tag{11.2}$$

where Q_t is the amount of drug dissolved in time t, Q_0 is the initial amount of drug in solution, and k_1 is the constant release of first order.

Higuchi model: This release pattern shows the amount of drug released according to the square root of time in which the dissolution took place according to the relation

$$Q_t = k_h \cdot t^{1/2} \tag{11.3}$$

where Q_t is the amount of drug dissolved in time t, k_h is Higuchi dissolution constant that depends on the matrix system variables, and t is the release time in hours.

Korsmeyer–Peppas model: Korsmeyer and Peppas developed a simple, semiempirical model that exponentially relates the drug release to the fractional release of the drugs. Hence, the final equation can be written as shown in relation (11.4)

$$M_t / M_0 = k \cdot t^n \tag{11.4}$$

where n is the diffusion exponent, k is a constant depending on the geometric and structural characteristics of the polymer matrix, n is the characteristic exponent of the diffusion transport type, and M_t/M_0 is the fraction of drug substance released at time t. This model allows defining the type of release mechanism regardless of the system's geometry. The diffusion exponent values (n) (Siepmann and Peppas 2001) depend on the various geometries in the polymer matrices and they are presented in Table 11.1.

Hixson–Crowell model: Hixson and Crowell proposed a correlation between drug release from the particle and diameter area of the particle. The proposed equation shows that the drug release from the particle is proportional to the cubic root of its volume, and can be described as shown in relation (11.5)

$$Q_0^{1/3} - Q_t^{1/3} = k_{HC} \cdot t \tag{11.5}$$

where Q_0 is the initial amount of drug in the polymer matrix, Q_t is the amount remaining after the release of the drug at time t, and k_{HC} is Hixson–Crowell release constant which depends on the surface and volume properties of the polymer matrix.

TABLE 11.1
Diffusion Exponent Values (n) Depending on the
Geometry of the System

Diffusion Exponent (n)			
Film	Cylinder	Sphere	Release Mechanism
0.5	0.45	0.43	Fick diffusion
$0.5 < n < 1$	$0.45 < n < 0.89$	$0.43 < n < 0.85$	Abnormal transport
1	0.89	0.43	Type II transport

11.4 ANALYSIS OF DONEPEZIL RELEASE FOR POLYMORPH A AND POLYMORPH B AT DIFFERENT SIZES OF THE PARTICLES

The controlled release is determined by the polymer structure and its features (Liew et al. 2012). The release of donepezil concentration from the patch is shown on the calibration curve. To obtain the calibration graph, 100 mg of donepezil was melted in 50 mL water, then heated at 80°C and mixed up till all donepezil quantity introduced in the pH = 5.5 tampon solution is dissolved, and it is obtained in a *clear solution*. The *clear solution* will indicate the point from the calibration graph corresponding to 100% concentration. After that, the solution is diluted in order to obtain the other points on the calibration graph (Zelinschi and Dorohoi 2012). The dilution is done as follows:

- One volume of 6.6 mL *clear solution* is completed with 3.3 mL tampon solution and the electronic absorption spectrum is recorded. The intensity in maximum of the electronic absorption band and the concentration of 66% will represent another point on the standard graph.
- A volume of 5 mL of *clear solution* is completed with 5 mL tampon solution, and the value of the intensity in maximum of the electronic absorption band and the concentration of 50% will be the coordinates of another point on the calibration curve.
- A volume of 3.3 mL of *clear solution* is completed with 6.6 mL tampon solution, being obtained by the same procedure; the point on the calibration graph corresponds to the concentration of 33%.

It should be noted that the experiments used for a transdermal-controlled release can be calibrated using any of the polymorphs of the drug; since the substance once dissolved in the buffer solution, one cannot speak about the crystal structure or an arrangement of molecules in the structure of donepezil. Table 11.2 shows the concentrations of the active substance–donepezil corresponding to the absorbance $\lambda = 232$ nm.

The donepezil calibration curve is shown in Figure 11.1. By fitting (soft Origin Pro 8), a straight line was obtained with equation $y = 0.032x - 0.167$ where x is the concentration of donepezil in solution and y is the absorbance at $\lambda = 232$ nm. The correlation coefficient is $R = 0.995$.

TABLE 11.2

Concentrations of the Active Substance–Donepezil Corresponding to the Absorbance λ = 232 nm

Donepezil Concentration (%)	Absorbance (au)
100	3.025
66	1.974
50	1.522
33	0.828

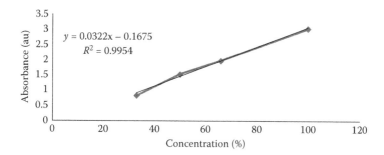

FIGURE 11.1 The calibration graph for donepezil in buffer solution with pH = 5.5. (Reprinted with permission from Zelinschi, B.C. and Dorohoi, D.O. 2012. *Rev. Chim. (Bucharest)* 63(8):811–814.)

11.4.1 Donepezil (BPMDMDHI)–Polymorph A

The controlled-release graphs of donepezil–polymorph A have been carried out starting from the experimental determinations of the two different dimensions of the active substance particles. The profile of the donepezil (BPMDMDHI) polymorph A—controlled release included in PVA matrix for the two different sizes of the particles—30 and 100 μm is shown in Figure 11.2.

Donepezil polymorph A with dimensions of 30 μm was faster released compared to that having dimensions of 100 μm. The first type of donepezil (BPMDMDHI) polymorph A—30 μm was released in the first 15 min in a ratio of about 69%, while for the donepezil–polymorph A—100 μm in the first 15 min was released at only 23%. For both types of particles after the first 8 h, the concentration of the released donepezil–polymorph A is relatively constant. The obtained experimental data have been worked out and analyzed from the perspective of application of the controlled-release kinetics mathematical models. In Figures 11.3 through 11.5, the graphs are shown corresponding to the maximum values of the correlation coefficients obtained for the models of the controlled-release kinetics at both the samples of donepezil polymorph A included in the PVA matrix.

The values of the correlation coefficients (R^2) and dissolving constants (K) as well as the constant which dictates the transport phenomenon (n) corresponding to the

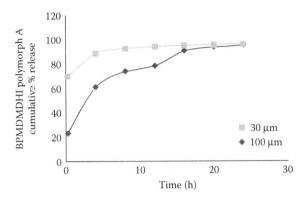

FIGURE 11.2   The profiles of the release of donepezil (BPMDMDHI)–polymorph A at a size of 30 µm and 100 µm of the particles in the PVA polymeric matrix. (Reprinted with permission from Zelinschi, B.C. and Dorohoi, D.O. 2012. *Rev. Chim. (Bucharest)* 63(8):811–814.)

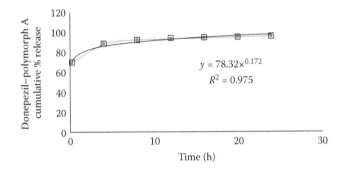

FIGURE 11.3   Korsmeyer–Peppas kinetic, PVA and donepezil polymorph A (30 µm). (Reprinted with permission from Zelinschi, B.C. and Dorohoi, D.O. 2012. *Rev. Chim. (Bucharest)* 63(8):811–814.)

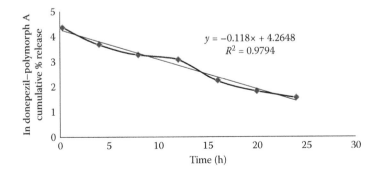

FIGURE 11.4   The first-order kinetic, PVA and donepezil polymorph A (100 µm). (Reprinted with permission from Zelinschi, B.C. and Dorohoi, D.O. 2012. *Rev. Chim. (Bucharest)* 63(8):811–814.)

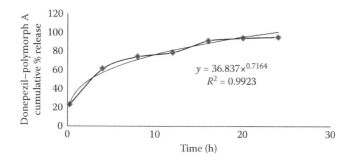

FIGURE 11.5 Korsmeyer–Peppas kinetic, PVA and donepezil polymorph A (100 μm). (Reprinted with permission from Zelinschi, B.C. and Dorohoi, D.O. 2012. *Rev. Chim. (Bucharest)* 63(8):811–814.)

release kinetic models (Park et al. 2007) for the two donepezil polymorph A samples (30 mm and 100 μm) are shown in Table 11.3.

11.4.2 Donepezil (BPMDMDHI)–Polymorph B

In Figure 11.6, it is shown that donepezil polymorph B is included in the PVA matrix with particles of 30 μm size controlled-release profile.

For donepezil–polymorph B with 30 μm size of the particles, the obtained experimental data have been worked out and analyzed from the perspective of application of all controlled-release mathematical kinetics models known from the literature. One can note that if the model for which one has the highest correlation coefficient-determined value is the zero-order kinetic model, then, for the 30 μm size of

---

TABLE 11.3

Values of the Correlation Coefficient and Dissolving Constants of the PVA Samples with Donepezil–Polymorph A for Different Study Models of the Release Kinetic

Sample	Zero-Order Kinetic		First-Order Kinetic		Higuchi Model		Korsmeyer–Peppas Model			Hixson Model	
	K_0	R^2	K_1	R^2	K_h	R^2	K	R^2	n	K_{HC}	R^2
PVA+D polymorph A (30 μm)	0.851	0.609	−0.076	0.832	5536	0.819	78,324	**0.975**	0.712	−0.005	0.829
PVA+D polymorph A (100 μm)	2681	0.810	−0.118	0.979	16,290	0.953	36,837	**0.992**	0.716	−0.059	0.712

Source: Data from Zelinschi, B.C. and Dorohoi, D.O. 2012. *Rev. Chim. (Bucharest)* 63(8):811–814.

Note: Bold values indicate that the results for this model (Korsmeyer–Peppas) are the best from all models involved.

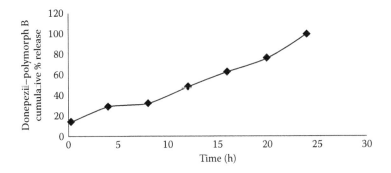

FIGURE 11.6 Donepezil–polymorph B release profile for a 30 μm particles size in the PVA. (Reprinted with permission from Zelinschi, B.C. and Dorohoi, D.O. 2012. *Rev. Chim. (Bucharest)* 63(8):811–814.)

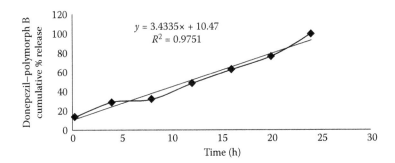

FIGURE 11.7 Zero kinetic order for donepezil–polymorph B (30 μm) + PVA. (Reprinted with permission from Zelinschi, B.C. and Dorohoi, D.O. 2012. *Rev. Chim. (Bucharest)* 63(8):811–814.)

donepezil–polymorph B, the best release kinetic is that corresponding to the zero-order kinetic model shown in Figure 11.7.

11.5 INDUCED BIREFRINGENCE AND THE INFLUENCE OF THE STRETCHING PROCESS ON THE DONEPEZIL RELEASE FROM THE PVA FOILS

The polymer foils are characterized by the birefringence which derives from the asymmetry of the molecular structures of the polymer. The asymmetry of the chemical structures of the polymer determines an intrinsic optical anisotropy of them caused by the different orientation of the chain units (Nechifor et al. 2010). If the polymer foils are stretched, a supplementary-induced birefringence is added to the initial intrinsic optical anisotropy. This induced birefringence when PVA foils are stretched can be measured using a Babinet Compensator and it is given by the difference between the two main refractive indices: n_e (λ)—the refractive index of the film for radiation having the electric field direction parallel to the stretching direction and

$n_o(\lambda)$—the refractive index of the film for radiation having the electric field direction perpendicular to the stretching direction. The induced birefringence when PVA foils are stretched is expressed by the relation

$$\Delta n(\lambda) = n_e(\lambda) - n_o(\lambda) \tag{11.6}$$

Solutions with various concentrations of donepezil were obtained by dissolving different amounts of donepezil in ethanol and it was incorporated afterward in a PVA solution of 10%. The mixture of PVA and donepezil solution was stirred at ambient temperature for 48 h. Homogenous and transparent PVA foils having a concentration of 2%, 3%, and 6% of donepezil were realized.

If the polymer films are stretched, their optical birefringence increases and it is considered as a measure for the orientation of the molecular chains of the foils. Because PVA films containing donepezil in different concentration investigated at various concentrations have approximately the same thickness (1 mm), from experimental data, one can observe an increased tendency of the birefringence by the stretching degree. The stretching degree of the PVA with donepezil of 2%, 3%, and 6% foils was evaluated by the ratio of the semiaxes of an ellipse in which a circle that drowns on the polymer film degenerates. One can calculate this stretching degree by the relation

$$\gamma = a/b \tag{11.7}$$

where a and b are the lengths of the large and respectively small semiaxes of the ellipse results after stretching. Experimental data were made only for small values of the stretching degree (for γ up to 1.4) because after that the foils have become rigid and could not be stretched more. For small degrees of stretching, one can observe a linear increase of the birefringence by the stretching degree (Figure 11.8).

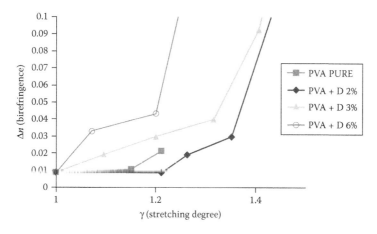

FIGURE 11.8 The birefringence of pure PVA and PVA with donepezil in different concentration versus stretching degree.

TABLE 11.4

Correlation Coefficient Values and Constants Dissolution of PVA with Donepezil 2% Foils (PVA+D 2%) for Different Study Designs of the Release Kinetics at Different Stretching Degrees

Samples	Stretching Degree	Zero-Order Kinetic		First-Order Kinetic		Korsmeyer–Peppas Model		Hixson–Crowell Model	
		K_0	R^2	K_1	R^2	R^2	N	K_{HC}	R^2
PVA+D 2%	1	0.0031	0.788	−0.0042	0.9029	**0.988**	0.4690	0.0048	0.7457
PVA+D 2%	1.17	0.0029	0.9801	−0.0007	0.9835	**0.989**	0.5329	0.0021	0.9641
PVA+D 2%	1.235	0.0018	0.9515	−0.0048	0.9631	**0.984**	0.5629	0.0009	0.7709
PVA+D 2%	1.4	0.0154	0.8272	−0.0042	0.8486	**0.877**	0.6326	0.1101	0.7807

Note: Bold values indicate that the results for this model (Korsmeyer–Peppas) are the best from all models involved.

From Figure 11.8, one can observe how donepezil improves the orientation of polymer chains through physical interactions with the functional groups of PVA. The birefringence of pure PVA films is lower than that of PVA foils with donepezil of 2%, 3%, and 6%; by increasing the concentration of donepezil in the PVA film, the birefringence increases. The known kinetics mathematical models of controlled-release foils were applied to the PVA with donepezil of 2% (PVA + D 2%) at various degrees of stretching, and the synthetic results are shown in Table 11.4.

One can estimate based on the values of correlation coefficients the optimal release kinetic

- If for the PVA foils embedded with donepezil of 2% concentration, the best kinetic model is Peppas–Korsmeyer model.
- For the unstretched and stretched film of PVA with donepezil of 2% concentration, the diffusion exponent values for each stretching degree were calculated, and the values are presented in Table 11.5.

The values of the diffusion exponent depend on the degree of stretching, as it can be seen from Table 11.5. In the case when the degree of stretching is $\gamma = 1$, the value of the diffusion exponent indicates a normal diffusion process. In all three cases ($\gamma = 1.17$, $\gamma = 1.235$, and $\gamma = 1.4$), the diffusion exponent values are between 0.5

TABLE 11.5

Diffusion Exponent for PVA with Donepezil 2% Concentration Foils in the Controlled-Release Process

Stretching degree	1	1.17	1.235	1.4
Diffusion exponent	0.469	0.5329	0.5629	0.6326

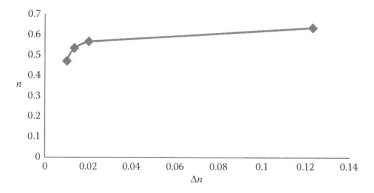

FIGURE 11.9 Variation of the diffusion exponent for PVA with donepezil 2% concentration foils versus birefringences.

and 1, indicating an abnormal transport or a non-Fickian transport, caused by phenomena such as hydration and swelling of the polymer matrix, the diffusion of the active substance, and/or erosion and degradation of the polymer matrix. In the case of stretched polymer films after hydration and swelling, the relaxation of polymer chains should take place. In this way, the drug release is delayed. This fact could explain the obtained values for n, in the range of 0.5–0.7 for the stretched foils, suggesting the overlapping of relaxation phenomenon upon the existing other phenomena. The total amount of the released donepezil from PVA with donepezil of 2% concentration foils increases with the stretching degree showing that the alignment of polymer chains can influence both the amount of drug delivered and the phenomena involved in the drug release process.

The variation of the diffusion exponent for PVA with donepezil of 2% concentration foils was studied by the birefringence and the dependence is shown in Figure 11.9.

The minimum value of the exponent of diffusion is 0.46 and corresponds to a birefringence of 0.012. The diffusion exponent increases by the increasing of the birefringence for PVA with donepezil of 2% concentration foils to a maximum of 0.63 corresponding to a value of 0.124 for birefringence.

11.6 CONCLUSIONS

- The controlled-release study of an active pharmaceutical substance from the PVA polymer matrix provides clear indications about the polymorph that needs to be used in order to obtain the best release profile for the patient.
- *In vitro* release study of donepezil embedded in the PVA polymer matrix provides accurate information about the particle size of the active substance included in the PVA polymer matrix. Particle size is a key element in achieving the desired therapeutic effect.
- There are certain diseases such as AD, for which embedding a pharmaceutically active substance under a patch form and not a classic tablet brings an undoubtedly benefit for the patient and gives him some independence both

from a medical standpoint and an improvement in his personal life and no longer a dependence on a rigorous schedule of drug administration.

- The study of "*in vitro*" release for a pharmaceutically active substance–donepezil for Alzheimer's included in the PVA matrix was carried out for two different polymorphs of donepezil (A and B) and for two different sizes of particles with polymorph A of donepezil and then for PVA with donepezil of 2% concentration-stretched foils.

- The active substance–donepezil (BPMDMDHI) is released by diffusion through the porous structure. The release kinetic models of donepezil included in the PVA matrix have been applied for two polymorphs of donepezil-A and B.

- Using mathematical models for the determination of donepezil–polymorph A release kinetic type included in the PVA polymeric matrix, it is noted that the release is done for both samples with different sizes of the particles according to a kinetic of the Korsmeyer–Peppas type and for donepezil polymorph A with the size of 100 μm, it is recorded with a very good value of the correlation coefficient corresponding to the first-order kinetic model. The mathematical model with the highest determined value of the correlation coefficient for donepezil–polymorph B of the 30 μm particles size is of zero kinetic order.

- The influence of the stretching process on donepezil release from PVA foils was investigated in this chapter too and elucidated the mechanisms involved in kinetic dynamics of drug delivery into the human body as a function of the stretching degree.

- The birefringence of pure PVA films is lower than that of PVA foils with donepezil in different concentrations (2%, 3%, and 6%).

- The birefringence of PVA foils with 2% donepezil increases with the stretching degree, being higher than the birefringence for the pure PVA films.

- The diffusion coefficient, n, increases with the stretching degree, keeping its values in the range corresponding to an anomalous drug transport from the polymer matrix.

- According to the experimental data, the released concentration of the active substance increases with the stretching degree of the PVA with donepezil of 2% and that has been submitted to the extension process.

- Polymer matrices that include in their structure active substances for which one can observe the controlled-release mechanism that began to be widely used in both the pharmaceutical industry and other fields; for example, in agriculture for formulation on chemical fertilizers or substances used in pest control, etc.

REFERENCES

Adin, I., Iustain, C., Arad, O., and Kaspi, J. 2006. New crystalline forms of donepezil base, EPO Patent EP1669349, Chemagis, Justia Patents.

Lekshmi, U.M.D., Poovi, U.G., and Neelakanta Reddy, P. 2012. In-vitro observation of repaglinide engineered polymeric nanoparticles. *Dig. J. Nanomaterials Biostructures* 7(1):1–3.

Liew, K.B., Fung Tan, Y.T., and Peh, K.K. 2012. Characterization of oral disintegrating film containing donepezil for Alzheimer disease. *Am. Assoc. Pharm. Sci. PharmSciTech* 13(1):134–135.

Nechifor, C.D., Angheluta, E., and Dorohoi, D.O. 2010. Birefringence of etired poly (vinyl alcohol) (PVA) foils. *Rev. Materiale Plastice* 47:164–166.

Park, T.J., Ko, D.J., Kim, Y.J., and Kim, Y. 2007. Polymorphic characterization of pharmaceutical solids, donepezil hydrochloride, by CCP/MAS solid state nuclear magnetic resonance spectroscopy. *Bull. Korean Chem. Soc.* 30(9):123–124.

Rogojanu, A., Rusu, E., Olaru, N., Dobromir, M., and Dorohoi, D.O. 2011. Development and characterization of poly (vinyl alcohol) matrix for drug release. *Dig. J. Nanomaterials Biostructures* 6(2):809–810.

Rojas-Fernandez, C.H. 2001. Successful use of donepezil for the treatment of dementia with Lewy bodies. *Ann. Pharmacother.* 35(2):202–205.

Schroeder, I.Z., Franke, P., Schafer, U.F., and Lehf, C.M. 2007. Development and characterization of film forming polymeric solutions for skin drug delivery. *Eur. J. Pharm. Biopharm.* 65(1):111–112.

Siepmann, J. and Peppas, N.A. 2001. Preface: Mathematical modeling of controlled drug delivery. *Adv. Drug Deliv. Rev.* 48:137–138.

Speight, J.G. and Lange, N.A. 2005. *Lange's Handbook of Chemistry*, 16th edition. Maidenhead McGraw-Hill Professional, New York, 807.

Steele, L.S. and Glazier, R.H. 1999. Is donepezil effective for treating Alzheimer's disease? *Can. Fam. Physician* 45:917–919.

Sugimoto, H., Ogura, H., Arai, Y., Iimura, Y., and Yamanishi, Y. 2002. Research and development of donepezil hydrochloride, a new type of acetyl cholinesterase inhibitor. *Jpn. J. Pharmacol.* 89(1):7–20.

Zelinschi, B.C. and Dorohoi, D.O. 2012. In-vitro release study of (2-[(1-benzyl-4-piperidyl) methyl]-5,6-dimethoxy-2,3-dihydroinden-1-one) BPMDMDHI included in the PVA matrix. *Rev. Chim. (Bucharest)* 63(8):811–814.

12 Structuring of Polymer Surfaces via Laser Irradiation as a Tool for Micro- and Nanotechnologies

Iuliana Stoica and Nicolae Hurduc

CONTENTS

12.1 INTRODUCTION

Polymers are macromolecules, which are synthesized from one or more different monomers using different types of polymerization, namely radical or ionic polymerization, polycondensation, polyaddition, and copolymerization. The polymerization type has a direct influence on the characteristics of the polymer, such as molecular weight and distribution, impurities, and polymer structure. Also, the deposition method of the polymeric films has an important influence on surface characteristics (morphology, texture, and surface tension). Therefore, polymers can be chemically or physically tailored to be used in the desired applications: molecular electronics, biotechnology, optoelectronic devices, and liquid crystal (LC) display manufacturing, where the molecules alignment has significant importance. The alignment of the molecules can be induced by micrometric- or nanometric-grooved substrates varying in height, width, and spacing. There are many methods used to generate micro- and nanopatterned polymer surfaces, such as electron beam lithography, nanoimprint lithography, laser interference lithography and subsequent reactive ion etching, soft lithography, crack-induced grooving method, rubbing with different types of fabric, scanning probe lithography, and so on. Unfortunately, all mentioned techniques have two or more fabrication steps. A single-step processing technique able to create surface relief modulation preserving material properties is ultraviolet (UV) laser irradiation. The laser is one of the most widespread technical inventions of the last century. Since its discovery, laser polymer processing has become an important field of applied and fundamental research. There are many examples of common lasers used in polymer science, among them ArF excimer laser and Nd:YAG laser. The polymeric materials suitable for such a technique are those that have a photochromic behavior, namely azo-polymers. This type of polymer has the capacity to generate surface relief gratings (SRGs) as a consequence of light irradiation through an interference pattern. This behavior is due to the *trans–cis–trans* isomerization processes of the azobenzene groups, which are either linked to a polymeric chain, or part of azo-dye-doped polymers (guest–host systems). Considering the above, this chapter will present some essential issues regarding the interaction of the laser beam with the matter and the manner in which SRGs are generated with laser interference systems or interference masks through continuous or pulsed laser irradiation on azo-polymers.

12.2   BASIC CONCEPTS REGARDING THE POLYMERIC MATERIALS USED FOR SRG

As mentioned above, the polymeric materials adequate for fabrication of SRG are those that have a photochromic behavior, namely azo-polymers (Kim et al. 1995, Rochon et al. 1995, Petrova et al. 2003, Zucolotto et al. 2004). Owing to the *trans–cis–trans* isomerization processes of the azobenzene groups present in the polymer structure, these types of polymers have the capacity to generate SRGs as a consequence of light irradiation through an interference pattern. Viswanathan et al. classified azobenzene containing polymer systems into two groups (Figure 12.1), namely azo-dye-doped polymers and azo-dye-functionalized polymers, also divided into amorphous and liquid crystalline polymers.

The azo-dye-doped polymer systems, known as the guest–host system, are made by dissolving azo dyes together with the host polymer in an appropriate solvent. The concentration of dye in the polymer is adjusted to produce homogeneous films, formed from the solutions either by casting or by spin coating onto glass slides. Studies in this field have shown that the azo-dye-doped polymer films are not favorable to form SRGs, the surface relief features being very weak. This is due to the chromophores that are not tethered to the polymer chains and their photoisomeric movements are not hindered by the polymer chains (Viswanathan et al. 1999).

In azo-dye-functionalized polymer systems, the azobenzene chromophore is covalently linked either as a side chain to the polymer backbone or in the main chain of the polymer. Over time, a large variety of azo-functionalized amorphous polymers with different chemical structures, molecular weights, and glass transition temperatures (*Tg*) were synthesized and used to create SRGs: side-chain azo-polymers by direct polymerization, side-chain azo-polymers by postcoupling reaction,

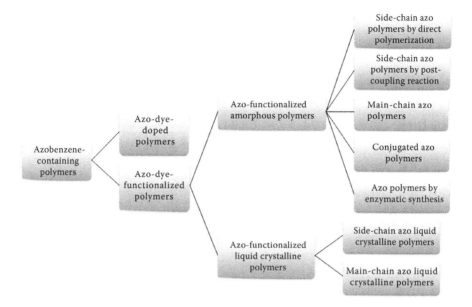

FIGURE 12.1 Classification of azobenzene-containing polymer systems used for SRGs fabrication.

main-chain azo-polymers, conjugated azo-polymers, and enzymatic synthesized azo-polymers.

The SRG formation was first studied on polymers with azobenzene chromophores in the side chain, synthesized by direct polymerization. The photoinduced surface deformation is the result of two processes that takes place in parallel: the first one is due to a special state of matter induced by the azobenzene groups that act as molecular motors and generate the polymeric chains displacement; the second one supposes an elastic deformation of the material due to the Weigert effect (Hurduc et al. 2016).

Post polymerization azo-coupling reaction can provide a simplest and convenient way to synthesize various kinds of azo-functionalized polymers with different degrees of functionalization on the same backbone structure. In addition, versatile functional groups can be incorporated on the azobenzene moiety. These azo-polymers show significantly different photoprocessability depending on the chromophore structures.

The SRG formation on main-chain azo-polymers, where the azobenzene groups are attached in the polymer backbones, was much slower than the side-chain polymers. This can be explained taking into consideration that the polymer has a rigid backbone and the azobenzene groups are bound to the backbone at both ends, restricting the mobility of the chromophores.

Recently, azo chromophores have been incorporated into polymers with π-conjugated backbone electronic structures. The SRG in this kind of azo-polymer systems, postfunctionalized with azobenzene groups, formed after the exposure of spin-coated films to the interference pattern, presents a very regularly spaced

surface structure and is expected to show interesting optical and electronic properties (Viswanathan et al. 1999).

Also, interesting photoanisotropic properties have been investigated and high-
efficiency SRGs were fabricated for novel photodynamic azo-polymers containing
azo chromophores synthesized by enzyme-catalyzed polymerization (Viswanathan
et al. 1999).

Liquid crystalline polymers with a pendant azobenzene group are due to linking
of the different mesogenic groups to the polymer main chain through flexible alkyl
spacers of varying lengths. Appropriately substituted azobenzene chromophores act
as mesogenic units as well. In the case of liquid crystalline polymers with azobenzene group in the side and main chain, the optically stored information is stable
below the Tg, has long-term stability if the Tg is higher than the ambient temperature,
and can be erased by heating the material above its Tg (Viswanathan et al. 1999).

Among the described classes of azo-polymers, our work group is focused mainly
on synthesis and characterization of polymers containing azobenzene groups in the
side chain.

12.3   IRRADIATION CONDITIONS USED FOR SINGLE-STEP
SURFACE RELIEF NANOSTRUCTURATION

Usually, SRGs can be generated with laser interference systems or interference
masks through continuous or pulsed laser irradiation, using a maximum light fluence lower than the ablation threshold of the material. The surface processing must
be made at wavelengths for which the polymers have a maximum of light absorption.

As a radiation source for pulsed laser irradiation experimental setup, an Nd:YAG
laser was used, working on its third harmonic at 355 nm, with a pulse length of
6 ns, a repetition rate of 10 Hz, and having a 0.6-mrad divergence (Damian et al.
2014). The laser energy was varied as needed and was permanently monitored with
an energy meter. The setup ensured that the fluence on the probe was far from the
ablation threshold of the polymers. To obtain the interference pattern, a 1000-nm or
500-nm diffraction-phase masks and only the 1 and 0 diffraction orders were used.
The irradiation scheme is presented in Figure 12.2. Because of the very short irradiation times (nanoseconds/picoseconds) used in the pulsed mode, a very fast chain
reorganization process can be expected, induced by the azobenzene dipole orientation because of the interaction of the laser-polarized light.

In the case of continuous laser irradiation, the polymer films were illuminated
with an interference pattern produced by the superposition of two coherent beams
incident symmetrically onto the film with respect to the film surface normal direction (Figure 12.3). The beam, delivered by a 488-nm-wavelength laser diode, was
incident to a beam splitter, and the resulting beams were superposed onto the film
after reflection onto two mirrors. The resulting interference pattern presented a sinusoidal modulation of the intensity along the polymer surface. The intensity in both
beams was adjusted with the optical densities being equal. The beam polarization
was set on the incidence plane. In this configuration, the polarization axes were
perpendicular to the fringes of the interference pattern, in order to favor a migration of the material from the high-intensity regions of the interference pattern to the

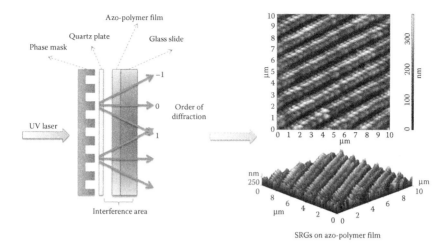

FIGURE 12.2 Pulsed laser irradiation experimental setup used to induce SRG on azo-polymer films. (Adapted from Stoica I et al. 2013. *Microsc. Res. Tech.* 76:914–923.)

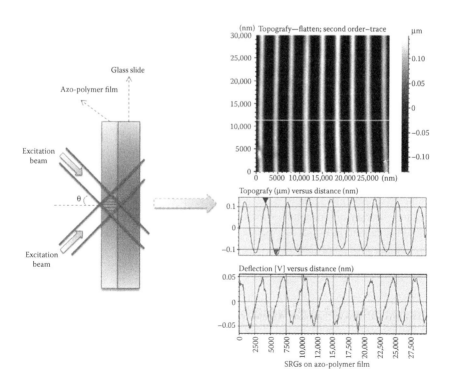

FIGURE 12.3 Continuous laser irradiation experimental setup used to induce SRG on azo-polymer films. (Adapted from Raicu Luca A et al. 2011. *eXPRESS Polym. Lett.* 5(11):959–969.)

low ones, even if the glass-transition temperature value of the polymer significantly exceeds the experimental conditions. A 633-nm-wavelength helium–neon laser was used to record the evolution of the first-order diffraction efficiency during the film illumination (Raicu Luca et al. 2011, Damian et al. 2014).

12.4   DEVELOPMENT OF THE NANOSTRUCTURED SURFACE RELIEF CREATED DURING LASER IRRADIATION PROCESS

In the last few years, several research groups reported the possibility to obtain an SRG by UV irradiation of polymers containing azobenzene. Among them, our group concerns both technical possibilities to obtain single-step surface relief modulation of the polymeric films in order to create integrated structures in materials (without any ablation and preserving material properties) and physicochemical processes responsible for single-step surface modulation under the action of light and time stability (Apostol et al. 2009, 2010).

Our studies in this field started with the investigation of some polymers obtained in a two-step reaction, starting from a polysiloxane containing chlorobenzyl groups in the side chain. In the first step, the polysiloxane was modified with 4-hydroxyazobenzene and in the second one, the unreacted chlorobenzyl groups were modified with some nucleobases, namely adenine and thymine. The obtained azo-polysiloxanes contain 60% azobenzene and around 20% nucleobases (Hurduc et al. 2007a,b). Depending on the chemical structure of the nucleobases, the surface of the azopolysiloxane films can be structured in different ways under UV irradiation. Also, if the UV irradiation experiment is done in dark, the azobenzenic groups change their configuration from *trans* to *cis* and maintain this new state more than 48 h after irradiation. Instead, taking into consideration that the UV light induces the *trans–cis* isomerization process and the visible light induces the reverse phenomenon, the *cis–trans* relaxation, if the UV irradiation is made in the presence of natural visible light, continuous isomerization processes *trans–cis–trans* take place. In the case of thymine-substituted azopolysiloxanes (when H-bonds are generated at supramolecular level), the conformational changes generated by the azobenzenic groups' photoisomerization, combined with the possibility of H-bond formation, induce a reorganization process at the film surface. Also, it can be observed that after UV irradiation, the surface roughness increases. If the adenine group is present in the side chain (when no H-bonds are formed), the surface roughness decreases after UV irradiation (Hurduc et al. 2007a,b).

Photoinduced single-step surface relief modulation of previously analyzed nucleobases containing azo-polysiloxanes was tested by Enea et al. Since the azopolysiloxane (AzoPsi) containing adenine presents some difficulties concerning the *cis–trans* relaxation process in the presence of natural visible light (behavior explained by complex formation between adenine- and azo-groups having a *cis*-configuration), the SRG capacity was only studied for the polysiloxanes modified with azophenols and thymine. Thus, it was possible to correlate the light-induced effects with a well-characterized material structure. Using the Nd:YAG laser harmonic of 355 nm, with pulse length of 6 ns and a phase mask of 1 μm pitch, an interference image with the period of the same order of magnitude as that of the phase mask was formed on the sample surface. The incident laser fluence and the number of laser pulses influence the nanostructuration. For low fluence (8.4 mJ/cm²) and low number of pulses, the structuration was not complete. Increasing the number of incident laser pulses to 100, a very uniform line structure was obtained. Using a fixed number of incident laser pulses and increasing incident laser fluence till 196 mJ/cm² (value situated in the laser ablation domain), the lines become narrower and the relief structuration is damaged. This fact indicates that the mechanism responsible for the SRGs formation is an inner material reorganization and not the material ablation. Also, these induced structures were stable in time at normal ambient temperature (Enea et al. 2008).

Sava et al. compared the photochromic behavior and surface-structuring capacity of azopolymers having flexible polysiloxane and rigid polyimide structures, respectively (Sava et al. 2009, 2013). The thymine-substituted AzoPsi was obtained as previously described (Hurduc et al. 2007a,b). The azo-polyimide (AzoPI) was synthesized using a two-step polycondensation reaction. The first step was performed with equimolar amounts of hexafluoro-isopropylidene-diphthalic-dianhydride and a chromophore-diamine, 2,4-diamino-4′-methylazobenzene, in *N*-methyl pyrrolidinone (NMP). In the second step, the chemical imidization of the obtained polyamic acid solution was performed (Sava et al. 2008). The photochromic studies evidenced that the azobenzene groups can isomerize in flexible polysiloxane case, but also in the case of the rigid AzoPI. Considering the ratio between the *trans–cis* isomerization and the *cis–trans* relaxation processes, both materials were recommended for SRG tests. The study regarding the surface-structuring capacity of these polymers showed that the irradiation conditions significantly influence the surface geometry of the polymers. For example, variation in the incident laser fluence was useful to control the modulation depth of the interference pattern (with pitch of the same order of magnitude as the diffraction phase mask one) from tens to hundreds of nanometers. Owing to the higher flexibility of the polysiloxane macromolecular chain as well as to the increased free volume of this amorphous polymer, a low-energy density (8.4 mJ/cm²) and a reduced number of pulses were enough to induce well-defined structured formations (Figure 12.4). In the case of the polyimide containing pendant azobenzene groups, the surface-structuring process is still possible in spite of the high rigidity of the polymer chain. Using the same numbers of irradiation pulses and increasing the incident fluence, the nanogrooves depth increased even tens of times. For a large number of pulses of irradiation (till 100) and incident fluence of 8.4 mJ/cm², the film was uniformly patterned, but when using 35 mJ/cm², the precision of the modulated relief was reduced only after 15 pulses of irradiation (Figure 12.5). The surface relief has the tendency to be erased after thermal treatment near the glass transition, reflecting that the structuring process was not accompanied by ablation phenomena.

Subsequently, for further applications, Stoica et al. evaluated the ordered and directionated nanostructures obtained on irradiated AzoPI films from the polar representation for texture analysis, by means of parameters such as isotropy and periodicity. According to the data exposed in Table 12.1, as the incident laser fluence and the number of pulses increase, the surface isotropy increases, due to the irregular surface structures that occur. For low-incident fluence of 8.4 mJ/cm², the surface root mean square roughness (Sq), which quantifies the shape–size complexity, increases as the

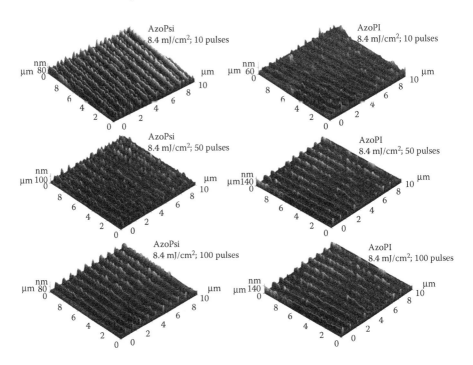

FIGURE 12.4 3D AFM images of the structured surface of azo-polysiloxane and azo-poly-imide films after irradiation with incident laser fluence of 8.4 mJ/cm^2 and 10, 50, and 100 pulses.

number of irradiation pulses increase. When using a high-incident fluence (35 mJ/cm^2), the highest value of Sq (106.2 nm) was found in the case of AzoPI$_{35/5}$ sample, for which the surface relief was uniformly patterned. Also, Sq tends to decrease as the disorder in morphology occurs (Table 12.1). On the other hand, the graphical studies of the functional volume parameters evidenced the ability of the modulated surfaces to carry out their function in a tribological contact, for different applications. Analyzing the values obtained from the Abbott–Firestone curves for the material volume parameters from Table 12.1, one can observe that using 35 mJ/cm^2 leads to a well-defined nanostructured surface relief. This was confirmed by higher values of peak material volume (Vmp) and core material volume (Vmc), indicating that more material can be worn away for a given depth of the bearing curve and more material can be available for load support once the top levels of the surfaces are worn away, compared to those obtained using 8.4 mJ/cm^2. Regarding the void volume parameters, the lower values of valley void volume (Vvv) indicate that the narrow pattern minimizes large areas that can trap gases or retain lubricants or remains of any kind, compared to core void volume (Vvc) values which suggest a higher void volume zone available to bear a load. Therefore, according to the desired application, a suitable irradiation condition that meets all the requirements regarding morphology, isotropy, periodicity, roughness, or functional properties will be selected (Stoica et al. 2013).

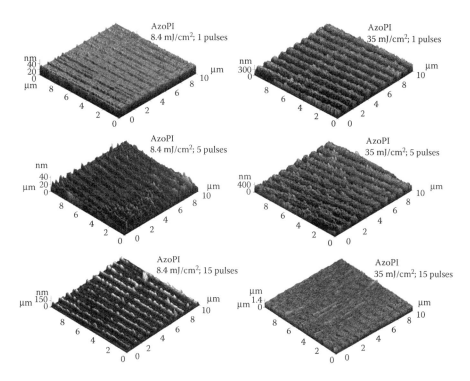

FIGURE 12.5 3D AFM images of the structured surface of azo-polyimide films after irradiation with incident laser fluence of 8.4 mJ/cm^2 and 35 mJ/cm^2 and 1, 5, and 15 pulses.

Resmerita et al. also investigated the capacity of an AzoPsi to generate nanostructured surfaces. The polymer was obtained in a two-step reaction, starting from a polysiloxane containing chlorobenzyl groups in the side chain. First, the polysiloxane was modified with 4-hydroxyazobenzene (68%–95% substitution degree) and then the unreacted chlorobenzyl groups were modified with donor/acceptor groups, namely azonaphthalene 93% (Kazmierski et al. 2004, Hurduc et al. 2007a,b). For irradiation, the third harmonic of an Nd:YAG laser with a wavelength value of 266 or 355 nm, incident fluence of 7.5 mJ/cm^2, pulse length of 6 ns, and 10 pulses was used (Resmerita et al. 2009). When the sample was irradiated with the wavelength of 266 nm, a linear-beaded periodical network with a modulation depth situated between 80 and 100 nm and a periodicity of 500 nm appear on the surface. Instead, when the 355 nm wavelength was used, well-defined periodical structures with a periodicity of 1000 nm and modulation depth of about 250–300 nm were obtained. Therefore, the obtained topographical features and relief depth can be controlled by different wavelengths and phase masks. Also, considering the very short irradiation time (10 pulses of 6 ns = 60 ns), the structuration mechanism was probable based on the instantaneous reorganization process, due to the dipoles orientation induced by the polarized light–Weigert effect (Resmerita et al. 2009).

TABLE 12.1

Surface Isotropy and Periodicity of the Nanogrooves, Roughness, and Functional Volume Parameters Obtained for the Irradiated Samples

Sample	Isotropy (%)	Periodicity (%)	Sq (nm)	Vmp (mL/m²)	Vmc (mL/m²)	Vvc (mL/m²)	Vvv (mL/m²)
AzoPI$_{8.4/1}$	2.32	24.5	3.4	$0.25 \cdot 10^{-3}$	$1.39 \cdot 10^{-3}$	$3.07 \cdot 10^{-3}$	$0.42 \cdot 10^{-3}$
AzoPI$_{8.4/5}$	1.67	62.8	5.7	$0.39 \cdot 10^{-3}$	$2.87 \cdot 10^{-3}$	$7.31 \cdot 10^{-3}$	$0.37 \cdot 10^{-3}$
AzoPI$_{8.4/10}$	1.55	54.1	11.5	$1.25 \cdot 10^{-3}$	$5.06 \cdot 10^{-3}$	$14.9 \cdot 10^{-3}$	$0.70 \cdot 10^{-3}$
AzoPI$_{8.4/15}$	2.92	60.4	28.6	$1.39 \cdot 10^{-3}$	$8.92 \cdot 10^{-3}$	$18.8 \cdot 10^{-3}$	$0.88 \cdot 10^{-3}$
AzoPI$_{8.4/50}$	2.54	52.7	14.1	$1.22 \cdot 10^{-3}$	$20.7 \cdot 10^{-3}$	$50.9 \cdot 10^{-3}$	$0.97 \cdot 10^{-3}$
AzoPI$_{8.4/100}$	2.80	63.1	29.9	$1.57 \cdot 10^{-3}$	$17.9 \cdot 10^{-3}$	$52.3 \cdot 10^{-3}$	$1.28 \cdot 10^{-3}$
AzoPI$_{35/1}$	3.34	78.7	97.6	$1.95 \cdot 10^{-3}$	$101 \cdot 10^{-3}$	$86.9 \cdot 10^{-3}$	$10.9 \cdot 10^{-3}$
AzoPI$_{35/5}$	3.73	71.8	106.2	$3.02 \cdot 10^{-3}$	$87.8 \cdot 10^{-3}$	$78.1 \cdot 10^{-3}$	$16.6 \cdot 10^{-3}$
AzoPI$_{35/10}$	5.11	71.2	90.2	$4.33 \cdot 10^{-3}$	$98.5 \cdot 10^{-3}$	$121 \cdot 10^{-3}$	$10.2 \cdot 10^{-3}$

Source: Data adapted from Stoica, I. et al. 2013. *Microsc. Res. Tech.* 76:914–923.

Note: The samples were named using the label $PI_{j/k}$, where j is the incident fluence and k is the number of irradiation pulses.

Recently, Sava et al. tested the structuring capacity of some azo-copolyimides free-standing films using two laser fluencies of 10 mJ/ cm² (Figure 12.6) and 45 mJ/ cm² (Figure 12.7) with 10 and 100 pulses number. The azo-copolyimides have been synthesized by the polycondensation reaction of hexafluoroisopropyli-dene-diphthalic-dianhydride and a mixture of two aromatic diamines, one of which contains ether groups, such as bis(p-aminophenoxy)-1,4-benzene (for AzoCPI1), bis(p-aminophenoxy)-1,3-benzene (for AzoCPI2), or bis(p-aminophenoxy)-4,4'-biphenyl (for AzoCPI3), and the other one contains a pendent-substituted azoben-zene group, namely 2,4-diamino-4'-methylazobenzene. The azo-copolyimide film which contains *meta*-connected linkages showed an irregular surface photoinduced pattern for 100 pulses and low-energy fluency and for high-energy fluency. This can be explained by slightly higher flexibility of the polymer. The photoinduced pat-tern was generated in this polymer, but the SRG could not be fixed well, because of the fairly high mobility of the polymer chain which tends effectively to relax to a smooth surface due to surface tension forces. Instead, the best results were obtained for azo-copolyimides with *para*-connected units by using high-density energy. The uniform SRG was obtained with a maximum photoinduced pattern depth of 240 nm for bis(p-aminophenoxy)-1,4-benzene containing the polymer by using an energy fluency of 45 mJ/cm² and 100 pulses. These results were also argued by the texture direction parameter (Table 12.2), which indicates in this case a good anisotropy of the morphology (Sava et al. 2015).

A parallel discussion between the characteristics of SRGs obtained by pulsed and by continuous-wave laser irradiation was made by Damian et al. on modified polysiloxane and poly(chloromethyl styrene) (Damian et al. 2014). The starting

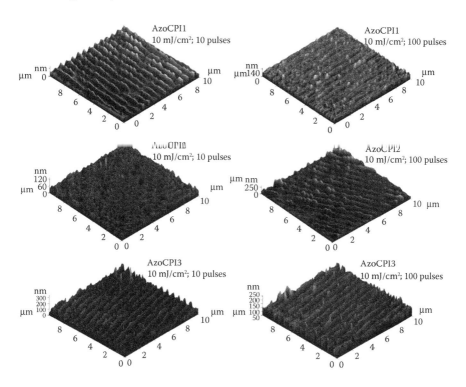

FIGURE 12.6 3D AFM images of the structured surface of azo-copolyimide films after irradiation with incident laser fluence of 10 mJ/cm² and 10 and 100 pulses.

polysiloxane, which contained chlorobenzyl groups in the side chain, was obtained using a two-step reaction, from dichloro(4-chloromethyl-phenylethyl)methylsilane. The polysiloxanes were then modified with different azophenols, namely 4-phenyl-azophenol, 4-(naphthalene-azo)-phenol, and 4-(4'-trifluoromethyl-phenylazo)-phenol, with an SN_2 reaction in dimethyl sulfoxide (Kazmierski et al. 2004, Hurduc et al. 2007a,b). Also, poly(chloromethyl styrene) was modified with 4-(phenyl-azo-phenol) under similar conditions, as in the case of the polysiloxane. The possibility of obtaining stable SRGs was tested on spin-coated films. When using pulsed laser irradiation (pulsed Nd:YAG laser, working on its third harmonic at 355 nm with a pulse length of 6 ns and a repetition rate of 10 Hz) on AzoPsi with polymeric chains grafted with unsubstituted azobenzene groups, surface relief instability was observed after 1 h after inscription, the SRG domains disappearing one by one. Instead, the SRGs remained stable if in the polysiloxanic chain, azonaphthalene or trifluoro-methyl (CF_3)-*para*-substituted azobenzene groups were connected. Also, in the case of pulsed laser irradiation, a very high sensitivity of the relief features as a function of the operational parameters (laser fluence and pulse number) was observed. When the continuous-wave laser irradiation was used, the polymer films were illuminated with an interference pattern produced by the superposition of two coherent beams (with an average intensity of 180 mW/cm²) resulted from splitting the beam delivered by a 488-nm-wavelength laser diode. In this case, the obtained SRGs were stable, even in

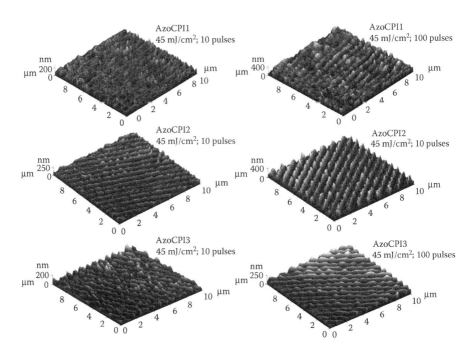

FIGURE 12.7   3D AFM images of the structured surface of azo-copolyimide films after irradiation with incident laser fluence of 45 mJ/cm^2 and 10 and 100 pulses.

TABLE 12.2

Texture Direction Parameter Calculated for the Structured Surface of Azo-Copolyimide Films after Irradiation

Sample	Stdi	Sample	Stdi	Sample	Stdi
AzoCPI1$_{10/10}$	0.303	AzoCPI2$_{10/10}$	0.529	AzoCPI3$_{10/10}$	0.232
AzoCPI1$_{10/100}$	0.546	AzoCPI2$_{10/100}$	0.295	AzoCPI3$_{10/100}$	0.252
AzoCPI1$_{45/10}$	0.571	AzoCPI2$_{45/10}$	0.197	AzoCPI3$_{45/10}$	0.261
AzoCPI1$_{45/100}$	0.437	AzoCPI2$_{45/100}$	0.211	AzoCPI3$_{45/100}$	0.195

the case of the polysiloxane substituted with azobenzene. Therefore, depending on the irradiation technique, different organization mechanisms can take place in the film (Damian et al. 2014). In the case of continuous laser irradiation, the SRG characteristics were strongly influenced by the laser wavelength. When using an argon–ion laser delivering a 355-nm-wavelength beam, the obtained SRG amplitudes were even 10 times lower compared to those obtained with a laser diode delivering a 488-nm-wavelength beam. The authors believed that in the case of the 488-nm-wavelength excitation, both the *trans* and *cis* forms of the azobenzene molecule were excited, favoring the relaxation of the azobenzene molecule from its metastable *cis* state to the fundamental *trans* state. Instead, in the case of the 355-nm-wavelength excitation, the *cis* isomer inside the material was in excess, this being unfavorable to the mass-displacement mechanism.

On flexible polysiloxane modified with 4-phenylazophenol, 4-((4′-nitrophenyl) azo)phenol, and 4-((4′-hydroxyphenyl)azo)benzonitrile and rigid poly(chloromethyl styrene) modified with 4-phenylazophenol, Hurduc et al. evidenced an interesting process consisting of surface relief erasure during SRG inscription, which results in reduction of the profile amplitude (Hurduc et al. 2014). The decreasing of the relief amplitude during the laser inscription is probable due to the presence of both elastic and plastic material deformations (Hurduc et al. 2016) as a consequence of the two parallel processes that take place during laser irradiation. The erasing phenomena are the result of the material relaxation of the elastic deformation component.

12.5 APPLICATIONS

The interest to elucidate the azo-polymers' nano-structuration is justified by their very recent applications in a variety of areas, such as optical information storage and processing, optical switching devices, diffractive optical elements, integrated optical devices (channel waveguides, polarization splitters, and nonlinear optical devices), holography, solar energy conversion, plasmonics (Hurduc et al. 2014), nanomanipulation (Hurduc et al. 2007a,b), biology, biotechnology, and medicine (Enea et al. 2008, Resmerita et al. 2009, Raicu Luca et al. 2011, Hurduc et al. 2013, 2014). Many of these applications are possible due to efficient photoisomerization and photoinduced anisotropy of the azobenzene groups.

In biology, in contrast to classical nanolithography requiring many preparation steps, the one-step SRG formed on azo-polymeric films can be used as a support for cell cultures, acting as a three-dimensional (3D) extracellular matrix (Hurduc et al. 2013).

The extracellular matrix geometry and rigidity can influence cell adhesion, with direct consequences on cell morphology and migration capacity, tissue architecture and regeneration, and stem cell fate/differentiation. The presence of the nucleobases together with azobenzenic groups in the polymeric side chains presents a potential interest for biological application, permitting to modify the surface structure (using physical H-bonding) in order to induce nanostructuration and thus, to favor the cell adhesion and cell directional development (Enea et al. 2008).

The capacity of the synthesized polymers having polysiloxanic main chain structure and different azobenzene groups in the side chain to generate nanostructured surfaces was tested by Resmerita et al., with respect to their potential application in biology, medicine, and biotechnology. The topography of the investigated biomaterials had a strong influence on cell function, adhesion, morphology, cytoskeletal arrangement, proliferation, and gene expression (Resmerita et al. 2009).

Promising results regarding the ability to support cell growth were also obtained for other types of azo-polysiloxanic films. Preliminary tests performed using the HepG2 and HeLa cell lines of human hepatic and epithelial origin showed remarkable properties to sustain both cell adhesion and proliferation. In addition, the films were easily sterilized by incubation in 100% ethanol and were stable during all cell culture procedures (Raicu Luca et al. 2011).

Another concern of the group was to study the interactions between the nano-structured surfaces of azo-polymers prepared starting from a polysiloxane containing chlorobenzyl groups and azo groups such as 4-(4'-(nitro-phenyl)azo) phenol, 4-((4'-(trifluoromethyl)phenyl)azo) phenol and 4-phenylazobenzene, and human hepatoma cells (HepaRG) (Hurduc et al. 2013). The chemical structure of the polysiloxanic films had a strong impact on cell–substrate interactions, influencing the division rate and adherence. The relief geometry had a strong effect on cell growth in the case of samples containing 4-(4'-(nitro-phenyl)azo) phenol, 4-((4'-(trifluoromethyl)phenyl)azo) phenol, and a higher number of cells being evidenced on the structured areas compared to the plane ones. The cytoskeleton structure showed a normal distribution of microtubules and actin filaments throughout the cytoplasm indicating a good adherence of the cells to these substrates. Instead, in the case of the third sample, the relief geometry had no influence on cell growth and did not improve the cell adherence properties. Both the plane and nano-structured surfaces accommodated a similar number of cells with a rather disorganized cytoskeleton (Hurduc et al. 2013).

Recently, Hurduc et al., investigating two types of azo-polymers having a flexible (polysiloxane) or rigid poly(chloromethylstyrene) main chain and chlorobenzyl groups and different azophenols in the side chain (4-phenylazophenol, 4-((4'-nitrophenyl)azo)phenol and 4-((4'-hydroxyphenyl)azo)benzonitrile), showed that the modification of Young's modulus value using light has a very interesting potential application in the cell culture field. By using different wavelengths, the flexibility of the AzoPsi cell substrate can be modified in a controlled manner, thus allowing the real-time investigation of the cell response right through changes in the mechanical properties of the extracellular matrix (Hurduc et al. 2014).

A unique opportunity, characteristic only to the azo-polymers, is represented by the possibility to generate dynamic relief surfaces, capable to change the relief characteristics in the presence of external stimuli, such as light or water (Rocha et al. 2014, Yadavalli et al. 2016). The interest for the dynamic surfaces is justified by the very recent studies concerning stem cells differentiation, that can be generated using only mechanical signals, induced by the extracellular matrix.

ACKNOWLEDGMENTS

This chapter was supported by a grant of the Romanian National Authority for Scientific Research and Innovation, CNCS–UEFISCDI, project PN-II-RU-TE-2014-4-2976, no. 256/1.10.2015.

REFERENCES

Apostol, I., Apostol, D., Damian, V., Iordache, I., Hurduc, N., Sava, I., Sacarescu, L., and Stoica, I. 2009. UV radiation induced surface modulation time evolution in polymeric materials. *Proc. SPIE* 7366:73661U-1-8.
Apostol, I., Hurduc, N., and Damian, V. 2010. Chapter 8. Tridimensional surface relief modulation of polymeric films. In *Polymer Thin Films*, ed. A.A. Hashim, 129–142. InTech, Rijeka.

Damian, V., Resmerita, E., Stoica, I., Ibanescu, C., Sacarescu, L., Rocha, L., and Hurduc, N. 2014. Surface relief gratings induced by pulsed laser irradiation in low glass-transition temperature azopolysiloxanes. *J. Appl. Polym. Sci.* 131:41015(1–10).

Enea, R., Hurduc, N., Apostol, I., Damian, V., Iordache, I., and Apostol, D. 2008. The capacity of nucleobases containing azo-polysiloxanes to generate a surface relief grating. *J. Optoelectron. Adv. Mater.* 10(3):541–544.

Hurduc, N., Ades, D., Belleney, J., Siove, A., and Sauvet, G. 2007a. Photoresponsive new polysiloxanes with 4-substituted azobenzene side-groups. Synthesis, characterization and kinetics of the reversible *trans*, *cis*, *trans* isomerization. *Macromol. Chem. Phys.* 208:2600–2610.

Hurduc, N., Donose, B.C., Macovei, A., Paius, C., Ibanescu, C., Scutaru, D., Hamel, M., Branza-Nichita, N., and Rocha, L. 2014. Direct observation of a thermal photofluidisation in azo-polymer films. *Soft Matter* 10:4640–4647.

Hurduc, N., Donose, B.C., Rocha, L., Ibanescu, C., and Scutaru, D. 2016. Azo-polymers photofluidisation—A transient state of matter emulated by molecular motors. *RSC Adv.* 6:27087–27093.

Hurduc, N., Enea, R., Scutaru, D., Sacarescu, L., Donose, B.C., and Nguyen, A.V. 2007b. Nucleobases modified azo-polysiloxanes, materials with potential application in biomolecules nanomanipulation. *J. Polym. Sci.: Part A: Polym. Chem.* 45:4240–4248.

Hurduc, N., Macovei, A., Paius, C., Raicu, A., Moleavin, I., Branza-Nichita, N., Hamel, M., and Rocha, L. 2013. Azo-polysiloxanes as new supports for cell cultures. *Mater. Sci. Eng.: Part C: Mater. Biol. Appl.* 33:2440–2445.

Kazmierski, K., Hurduc, N., Sauvet, G., and Chojnowski, J. 2004. Polysiloxanes with chlorobenzyl groups as precursors of new organic–silicone materials. *J. Polym. Sci.: Part A: Polym. Chem.* 42:1682–1692.

Kim, D.Y., Tripathy, S.K., Li, L., and Kumar, J. 1995. Laser-induced holographic surface relief gratings on nonlinear optical polymer films. *Appl. Phys. Lett.* 66:1166–1168.

Petrova, T.S., Mancheva, I., Nacheva, E., Tomova, N., Dragostinova, V., Todorov, T., and Nikolova, L. 2003. New azobenzene polymers for light controlled optical elements. *J. Mater. Sci.: Mater. Electron.* 14:823–824.

Raicu Luca, A., Rocha, L., Resmerita, A.-M., Macovei, A., Hamel, M., Macsim, A.-M., Nichita, N., and Hurduc, N. 2011. Rigid and flexible azopolymers modified with donor/acceptor groups. Synthesis and photochromic behavior. *eXPRESS Polym. Lett.* 5(11):959–969.

Resmerita, A.-M., Epure, L., Grama, S., Ibanescu, C., and Hurduc, N. 2009. Photochromic behaviour of nano-structurable azo-polysiloxanes with potential application in biology. *TOCBMJ* 2:91–98.

Rocha, L., Paius, C.-M., Luca-Raicu, A., Resmerita, E., Rusu, A., Moleavin, I.-A., Hamel, M., Branza-Nichita, N., and Hurduc, N. 2014. Azobenzene based polymers as photoactive supports and micellar structures for applications in biology. *J. Photochem. Photobiol. A: Chem.* 291:16–25.

Rochon, P., Batalla, E., and Natansohn, A. 1995. Optically induced surface gratings on azoaromatic polymer films. *Appl. Phys. Lett.* 66(2):136–138.

Sava, I., Burescu, A., Stoica, I., Musteata, V., Cristea, M., Mihaila, I., Pohoata, V., and Topala, I. 2015. Properties of some azo-copolyimide thin films used in the formation of photoinduced surface relief gratings. *RSC Adv.* 5:10125–10133.

Sava, I., Hurduc, N., Sacarescu, L., Apostol, I., and Damian, V. 2013. Study of the nanostructuration capacity of some azopolymers with rigid or flexible chains. *High Perform. Polym.* 25(1):13–24.

Sava, I., Resmerita, A.-M., Lisa, G., Damian, V., and Hurduc, N. 2008. Synthesis and photochromic behavior of new polyimides containing azobenzene side groups. *Polymer* 49:1475–1482.

Sava, I., Sacarescu, L., Stoica, I., Apostol, I., Damian, V., and Hurduc, N. 2009. Photochromic properties of polyimide and polysiloxane azopolymers. *Polym. Int.* 58:163–170.

Stoica, I., Epure, L., Sava, I., Damian, V., and Hurduc, N. 2013. An atomic force microscopy statistical analysis of laser-induced azo-polyimide periodic tridimensional nanogrooves. *Microsc. Res. Tech.* 76:914–923.

Viswanathan, N.K., Kim, D.Y., Bian, S., Williams, J., Liu, W., Li, L., Samuelson, L., Kumar, J., and Tripathy, S.K. 1999. Surface relief structures on azo-polymer films. *J. Mater. Chem.* 9:1941–1955.

Yadavalli, N.S., Loebner, S., Papke, T., Sava, E., Hurduc, N., and Santer, S. 2016. A comparative study of photoinduced deformation in azobenzene containing polymer films. *Soft Matter* 12:2593–2603.

Zucolotto, V., Barbosa Neto, N.M., Rodrigues Jr. J.J., Constantino, C.L., Zilio, S.C., Mendonsa, C.R., Aroca, R.F., and Olivera Jr. O.N. 2004. Photoinduced phenomena in layer-by-layer films of poly(allylamine hydrochloride) and Brilliant Yellow azodye. *J. Nanosci. Nanotechnol.* 4:855–860.

13 Liquid Crystal Polymers under Mechanical and Electromagnetic Fields
From Basic Concepts to Modern Technologies

Andreea Irina Barzic

CONTENTS

13.1 INTRODUCTION

Liquid crystals (LCs) represent a special category of materials with remarkable properties that arise from their molecular order that ranges between common liquids and solids. The self-aligning ability is also found in some macromolecular compounds, which exhibit similar mesophases characteristic of LCs, yet maintain certain useful features of polymers (Drzaic 1995, Cosutchi et al. 2010). One of the main characteristic of these compounds is the elongated shape of macromolecules and their intrinsic disposition toward alignment. However, polymers with LC character are able to form a regular lattice as in crystalline solids, but the strength of intermolecular forces is not of the same magnitude. In these conditions, "crystalline" character of liquid crystal polymers (LCPs) can be attributed to orientational order of the macromolecules, while their fluidity is determined by the lack of positional order (Brostow 1998). The liquid–solid duality generates the "anomalous" behavior of these organic materials. The unique dynamics of LCPs lies at the basis of creation of novel high-performance products with unique abilities.

Many efforts have been directed toward the development of molecular theories that allow a qualitative description of the molecular origin of various effects in LCPs (Coates 2000, Chen 2011). There are two aspects that should be examined in detail. The first one deals with the influence of basic characteristics of molecular structure, such as molecular shape and the polarizability anisotropy, on the physical features of LCPs. The second aspect concerns the sensitivity of certain properties to the details of molecular structure, namely flexibility and the positioning of particular elements of structure. The comparison of the theoretical and experimental data did not fully elucidate the connection between the features of the molecular structure and the macroscopic parameters of LCPs (Collyer 1992, Brostow 1998).

On the other hand, the properties of these soft and partially organized compounds can be adapted through application of external stimuli, such as electric, magnetic, and mechanical fields. In this context, new perspectives were opened not only in materials science, but also in optoelectronics and biomedicine (Scott et al. 2007, Cosutchi et al. 2011).

Having all these in view, this chapter attempts to describe the current state of art in the field of LCPs with tunable properties for advanced technologies. To understand the complex behavior of these materials, some basic concepts are presented, including the main categories of LCPs and their mesophases. The influence of some external factors on the LCPs physical characteristics is reviewed. The modifications induced by electric, magnetic, or mechanical shear fields on the morphology and optoelectronic properties of the LCPs are described. The most important applications of LCPs in various industries are briefly mentioned.

13.2   DEFINITION AND CLASSIFICATION

Some types of polymers exhibit LC properties, which are more complex than those of ordinary LC in that the internal ordering is influenced by additional factors related to the macromolecular structure (Collyer 1992, Shibaev and Lam 1994). These compounds can be classified in a similar manner as low-molecular-weight LC if considering the main aspects that trigger the self organization, namely temperature, concentration, and, in particular cases, the pressure. Thus, the three main categories of LCPs (depicted in Figure 13.1) are

- *Lyotropic LCPs*: the ordering of the macromolecules takes place by dissolving a polymer in an appropriate solvent. The phase diagrams of lyotropics reflect the essential influence of the compound concentration in the solvent at each temperature.
- *Thermotropic LCPs*: the orientation of polymer chains occurs when they are heated above the glass or melting transition point—the temperature being the only relevant factor in the phase diagrams.
- *Barotropic LCPs*: LC phase appears at elevated pressures. The pressure creates similar effects with those induced by temperature (but not identical) in terms of the volume modification.

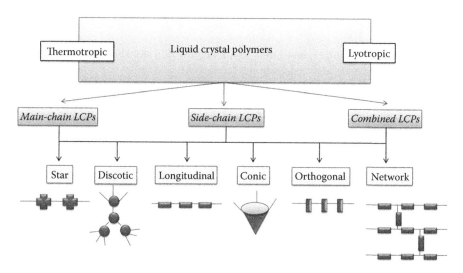

FIGURE 13.1   Classification of LCPs considering the ordering factors, the positioning of the mesogens and their shape.

The structural elements that render LC features to polymers are called *mesogens*. They can be inserted in distinct sites of the chains, leading to the following types of LCPs (see Figure 13.1):

- *Main-chain LCPs*: are obtained when the mesogens are inserted in the main chain of a polymer
- *Side-chain LCPs*: the mesogens are linked as side chains to the polymer backbone by a flexible connector named *spacer*
- *Combined LCPs*: are the result of combination of the main-chain and side-chain types

Starting from the aspects derived from positioning of the mesogens in the polymer structure, numerous and distinct macromolecular architectures have been reported, particularly liquid crystalline elastomers, dendritic LCPs, or block copolymers (Thakur and Kessler 2016). Supplementary factors are known to affect the meso-morphic features of polymers, namely molecular weight, length of flexible spacers, and regularity of succession of rigid and flexible units along the main chain (Thakur and Kessler 2016).

The shape of the mesogens represents another element against which LCPs may be classified (Scharf 2007, Dierking 2014). According to Figure 13.1, there are at least six types of LCP materials, namely

- *Longitudinal LCPs*: the mesogens are introduced in the main chain of linear polymers
- *Orthogonal LCPs*: the mesogen units are placed almost perpendicular to the backbone

- *Conic molecules of LCPs*: are organized polymers with three-dimensional character; all these aspects should render interesting electrical properties
- *Star LCPs*: are macromolecular compounds with a star-shaped conformation, which generates enantiotropic properties
- *Discotic LCPs*: are structures with diminished molecular mobility; they are further divided into three subcategories:
  - Single disks in the main chain and soft spacers between them
  - Single disks in the backbone but with rigid spacers
  - Multiple disks in the backbone center
- *LCP networks*: are compounds with elastomeric features that are prepared from LCPs containing reactive double bonds in the spacer sequences

Another classification can be made if one takes into consideration the inherent polarization characteristics of LCPs (Dierking 2014). On the basis of these properties, LCPs can be divided into two classes. The first one is represented by *ferroelectric LCPs*, which are a category of materials exhibiting a permanent polarization even in the absence of an electric field. Rod-shaped molecules in a specific phase exhibit both translational and orientational order. The mesogens arranged after a fixed spacing along one axis are able to form layers of oriented chains. In each planar layer, the director is tilted at an angle (sensitive to the temperature) from the normal. When chirality is present in the backbone, a gradual modification of tilt direction is noticed in successive layers of LC. Thus, the director precesses around the axis in adjacent layers, making with its surface a presumptive cone. A spontaneous polarization of molecules is formed because of the helical structure that occurred in the chiral mesophase. Therefore, all possible directions for the vector are tangent to the circle of intersection of the cone with the plane. The second one concerns *antiferroelectric LCPs*, where the director is tilted in the reverse direction and instant polarization in the opposite direction is noticed. Even so, the director still precesses around the axis; the distance along which the director precesses 180° (and not 360° as for ferroelectrics) denotes the pitch. This aspect is caused by the reverse tilt in neighboring layers, where the LC director goes around half of the cone. The director is found in the layer plane and the polarization vector is orthogonal to it. From one layer to another, the director is oriented in adverse directions, similarly to polarization vectors. In this way, the polarization vectors pointing up are equal to those pointing down. Thus, the spontaneous polarization is close to zero even for the nonhelical state.

13.3 PHYSICAL PROPERTIES UNDER THE INFLUENCE OF EXTERNAL FACTORS

In the presence of an external perturbation, such as shear, magnetic, or electric fields, one can distinguish major changes in the macroscopic properties of the LCPs (Drzaic 1995). The implementation of these fields mainly modifies the orientation of LC director. Consequently, in case of materials based on LCPs, the response to the action of the electromagnetic radiations or mechanical forces is different because the behavior is dictated by the configuration and conformation of polymer chains.

Among the most investigated properties in several industrial sectors, one can mention the degree of orientation of LC molecules and their speed of alignment. The latter could be changed according to the demands of the pursued application through the frequency and magnitude of the external field or through the intensity of mechanical deformation (Brostow 1998).

The physical characteristics of macromolecular compounds, particularly those of lyotropic systems, are significantly modified during shearing or extensional flow (Brostow 1998). The mechanical deformation determines an enhancement of the ordering degree of polymer chains. Furthermore, for structures with a certain degree of rigidity, one can observe peculiar textures when placing the sample under a polarized light microscope. These morphologies can be related to special optical properties and are mainly controlled by the nature of the solvent and the intensity of shear field.

The organization of macromolecules in LCP systems under electromagnetic waves strongly affects the dielectric constant and magnetic susceptibility (Coates 2000). As a result, they present distinct values along the LC director than along transverse directions. In this manner, the LC director is coupled with electric and magnetic fields contributing to the free energy. Equations 13.1 and 13.2 express previously mentioned aspects, but the order parameter imposed by the field is not taken into consideration in these relations

$$g_e = -\frac{\Delta\varepsilon}{8\pi} n \cdot E \tag{13.1}$$

$$g_z = -\frac{\Delta\chi}{2} n \cdot H \tag{13.2}$$

where ε is the dielectric constant, χ is the magnetic susceptibility, and n is the director of the LC, whereas E and H represent the intensity of the electric and magnetic field, respectively.

The molecular reorientation arises from the interaction of the field with the dipole moment of the polymer and it is not related with the conduction features of LCP. Taking into account the anisotropy of the physical properties described by Equations 13.1 and 13.2, the free energy of an LC in an external field shows a minimum for a certain orientation of the director in regard with the electromagnetic radiation. A positive anisotropy makes the director to align along the field, while for a negative anisotropy, the director stays orthogonal to the field (Brostow 1998). The orientation process is characterized by a threshold; its appearance is influenced by the condition where the torque of the electrical or magnetic forces is contrabalanced by the elastic forces.

Another important effect that has been evidenced in LCP compounds is the so-called Fréedericksz transition (Ma and Yang 2000). It can be defined as a transition from a uniform director configuration to a deformed director configuration—a change that is triggered in the presence of a sufficiently strong magnetic or electric field. This is not a phase transition, because at any point in the LCP, the extent of orientation of the macromolecules (relative to one another) remains the same. If the intensity of the field is not high enough, the director continues to be undistorted. As the field value is gradually increased from this limit point, the director starts to twist up to the point where it can follow the field. Thus, the Fréedericksz transition can take place in three different configurations (Kossyrev et al. 2002), known as the twist, bend, and splay geometries.

The main effects of the shear fields, magnetic, and electric are described in the following sections of the chapter.

13.3.1 SHEAR FIELD OR EXTENSIONAL FLOW

Application of shear or extensional flow to an LCP can lead to orientation of macromolecular chains. This influences the viscoelastic, morphological, and also other physical properties. In case of nematic longitudinal LCPs, the orientation is more efficient if the material is subjected to extensional flow rather than shear flow. In addition, cold drawing of such compounds determines a slippage of domains past each other without any remarkable improvement of the degree of orientation. Oscillatory shear tests provide insightful information on arrangements of LCP chains. An extensive investigation revealed that nematic, longitudinal LCPs derived from random copolymers such as those of chlorophenylene terephthalate and bis-phenoxyethane carboxylate can yield various textures (Graziano and Mackley 1984). The polarized optical microscopy studies showed that the morphology is affected by the molar mass of the copolymer. For samples with low molar mass, the microscopy images illustrate thin (1 μm) and thicker filaments at zero and low shear rates, respectively. The observed threads formed especially closed loops. Threads which were not connected to the surfaces are not fixed and they are able to move toward the "bulk" of the LCP film. The threads that were attached to the reverse surface were broken through shearing into dark lines, which become elongated along the external field direction. The field was composed of several stretched black lines on prolonged mechanical deformation. A gradual increase of the shear rate makes the texture of dark lines to be decomposed into short and curled entities (morphology named "worm texture") (Donald et al. 1983). During their flow along the shear field, the worms begin to modify their shape and they start to multiply as the shear rate is increased. Further enhancement of shear rate leads to formation of an ordered texture. The latter is evidenced by extinguishing incident radiation when putting the polarizer–analyzer pair parallel or at an angle of 90° to the flow direction. This oriented texture was characterized by fast relaxation in the first 10 s, beginning by occurrence of thin threads (close loops). After several minutes, the shrinkage and disappearance of the thin threads is observed. The quiescent sample with high molecular weight presents the worm texture and no thin threads are viewed. Shear conducted to good organization of chains along the deformation direction. The relaxation of the induced texture determined the formation of bands positioned orthogonal to the prior flow direction. It was presumed that in a shear gradient, molecules with reduced dimension tend to slip past each other and thus are able to orient along the flow direction. The aspect was noticed in the decrease of molecule viscous drag. For samples with higher molar mass, the independent movement of the molecules starts to be impeded by their length. In this way, the molecules are no longer able to flow so easily and they are constrained to restructure in the shear field. Structural modifications induced by shear should take place in a facile manner in LCPs with high molar mass.

Polydomains prior to shearing were evidenced in poly(*p*-hydroxybenzoic acid-co-2,6-hydroxynaphthoic acid) with 73 mol% *p*-hydroxybenzoic acid and poly(*p*-hydroxybenzoic acid-co-ethylene terephthalate) with 60 mol% *p*-hydroxybenzoic acid (Brostow 1998). The initial morphology consisted of sharp schlieren textures with disclinations, the latter starting to be multiplied after a critical shear rate. The multiplication process progressed to a point where the structures sizes are no longer in the range of the microscope resolution. At high shear rates, uniformly birefringent texture is noticed as a result of the breakdown of disclinations rather than the appearance of new ones (Kulichikhin 1988). At reduced shear rates, the polydomain morphology is assumed to be viscous owing to several disclinations which are supposed to be "crosslinks."

An important difference in regard to low molecular LCs is the formation of the so-called banded structures in LCPs. Many reports reveal the occurrence of the banded structure in a large number of longitudinal LCPs, both rigid and semiflexible, thermotropic and lyotropic subjected to a shear field (Donald et al. 1983, Navard 1986, Cosutchi et al. 2011, Barzic et al. 2015). It is shown that this type of induced morphology is universal in the appropriate domain of shear rate and viscosity, and is influenced by molar mass, temperature, and solvent type for lyotropic systems. In most cases, the bands are orthogonal to the shear direction and they are formed after the removal of shear force. The cause of such texture is presumed to be the result of a relaxation process. In particular conditions, they can form during the flow process itself or during elongation relaxation. The bands are ascribed to a periodical change of molecular director orientation in regard with the flow axis.

Literature (Ding et al. 1995) presents two possible periodic supermolecular structures of band textures for sheared LCPs, namely the sinusoidal model and the zigzag one. The banding in thermotropic copolyesters can be attributed to a serpentine trajectory of the chains (Brostow 1998). Many efforts have been made to explain the occurrence of shear bands based on flow instabilities and negative normal stresses (Marrucci and Greco 1993). For instance, poly(*p*-hydroxybenzoic acid-co-ethylene terephthalate) with 60 mol% *p*-hydroxybenzoic acid subjected to heating and shearing in a plate-to-plate rheometer presents banded morphology. However, the values of Reynolds number for this system were too small to generate flow instabilities. The explanation could be given by considering the theory proposed by Onogi and Asada (1990), which assumes that flow properties may be separated into three distinct regimes. At very low shear rates, the domains start to flow with significant rotation and slippage. Further enhancement of shear rate determines the deformation of the domains and at strong shearing, they are converted into a monodomain structure. For most LCPs, the band structure is distinguished in the low-to-intermediate shear rate regime. Here, two types of flow properties are noted: (1) the shear-thinning domain—characterized by distortional elasticity and (2) the Newtonian behavior reflecting a "dispersed polydomain" structure. After the removal of shear field, one may observe the relaxation behavior of the band structures and of the domains. The processes are similar in that both relax faster especially when the material is intensively sheared. This result supports the fact that the band spacing is reduced with application of a higher shear rate. Thus, the monodomain structure forms when LCP is subjected to very strong shearing.

The above-described aspects were evidenced for lyotropic solution of hydroxypropyl cellulose in specific solvents, such as water, acetic acid, methanol, or dimethylacetamide (Werbowyj and Gray 1984, Navard 1986, Cosutchi et al. 2010). For establishing the concentration at which the LC character appears, the viscosity of the polymer solution was registered over a wide range of concentrations. The maximum value of viscosity can be ascribed to isotropic–anisotropic transition, which takes place at about 38–45 wt%. The flow curves obtained for hydroxypropyl cellulose solutions in the lyotropic phase reveal a small Newtonian regime, followed by a shear-thinning domain, as expected for LCPs (Cosutchi et al. 2010).

The area of one-comb LCPs contains few investigations on rheological data that explain the shearing alignment. It was revealed that the flexible backbone strongly affects the rheological properties of the sample, suppressing any important manifestation of the LC feature of the polymers (Fabre and Veyssie 1987). Also, some nematic one-comb LCPs develop during flow that is a limited orientation of macromolecules close to the clearing point. The alignment in these systems arises from the coupling of the mesogens to the main chain.

Blends obtained from longitudinal LCPs and polymers with flexible backbones have been the subject of many works (Dutta et al. 1990). The system counterparts tend to phase separate and the material becomes aligned during the processing, creating the reinforcing (sometimes fibrous) component. The good degree of processability is ensured by the low viscosity of the nematic LCP components.

An important aspect of the above-presented band textures of LCPs is that they can be induced to isotropic polymers (Cosutchi et al. 2010, 2011, Barzic et al. 2015). The stability of the induced texture significantly depends on the interacting abilities of the blended polymers. In addition, if the isotropic polymer (polyamic acid or polyimide) can undergo cross-linking under electromagnetic radiation, the surface structuring generated by the LCP matrix is hardened. Immersion of the blend films at the composition at which the texture is more pronounced in a nonsolvent for the isotropic polymer does not remove the imprint given by the lyotropic matrix. Deeper investigations were focused on the influence of the dianhydride or diamine moieties on the rheological and morphological characteristics (Barzic et al. 2015). Two types of dianhydrides were used in isotropic polymer synthesis. The first structure is flexible and nonsymmetric (also nonplanar) and the second one presents a rigid and planar conformation. The strong anisotropy of some bands absorbance is supported by infrared dichroism data. The bands are wider as the amount of LCP in the system is lower, whereas their height evolution may be ascribed to a synergistic effect between flexibility and symmetry of the isotropic polymer. It was shown that the polymer derived from small, symmetric, and rigid dianhydride moieties is not able to fold on the LCP matrix and develops thicker bands and is slightly more irregular. Conversely, the polymer containing the large, nonsymmetric, and flexible dianhydride moieties is able to fold better on the periodic distribution of the LCP chains. Thus, it can interact more uniformly with them, generating to a finer band structure.

Other unpublished results reveal that the diamine residues present a great impact on the rheological and morphological properties of such blends containing LCP. The flow behavior of the systems containing cellulosic LCP and isotropic polymer is very complex, showing the mutual influence of the counterparts. The isotropic samples present a Newtonian behavior regardless of the applied shear rate. In combination with the LCP, the flow curves reflect the intermediate behavior of the two polymers in solution. Furthermore, depending on the diamine moieties structure (number of phenyl groups and ether bridges), the size of birefringent domains and the length of Newtonian regime can differ. For the isotropic imidic polymer containing more ether bridges in the neighboring of phenyl rings, the observed texture under polarized light consists of smaller brirefringent domains and the Newtonian plateau is longer when comparing with the sample with less ether linkages in the diamine residues. Examination of the morphology at nanometric scale reveals that blends based on imidic polymers in combination with a cellulosic LCP present a tendency of surface structuring, which is useful for display applications.

13.3.2 Electric Field

Application of electric fields to LCPs determines orientation of the molecular dipoles. The main difference from the low-molecular-weight LCs is that polymeric LCs present longer response times (Brostow 1998). Low-magnitude fields do not distort the director, while, as the field is increased from this point limit, the director starts to twist until it is aligned along the field. The field generates a slight rearrangement of electrons and protons in molecules, resulting in an induced electric dipole. Even if the alignment is not as strong as in the case of permanent dipoles, alignment with the electric field is still noticed. The induced orientation of the macromolecules is reflected in the anisotropy of the physical and morphological properties.

When the dipoles are "connected" to the mesogens, the latter can be aligned along a common director. Molecules with centrosymmetrical features are able to develop a quadrupolar orientation, while the substances without such features can form a structure with dipolar orientation in the presence of an electric field (Donald and Windle 1992). Generally, it can be stated that, in the presence of an electric field, longitudinal polymers are not so easy to orient as one-comb LCPs.

The main effect of this factor is that it serves as a removal of the domain boundaries by bringing the values of order parameters closer corresponding to the entire material and that of the monodomain. When too high voltages are applied, electrical degradation of the material takes place prior to alignment of LCP with longitudinal chains. The low-frequency conductance of LCPs is another factor that impedes the orientation in AC or DC fields. In some conditions, electric fields can affect the solidification of LCP materials (Martin and Stupp 1987). They can exhibit a preferential backbone orientation after isothermal aging of the mesophase and resolidification. Mesophase aging is also found in solid polymer samples with higher structural order. The initial main-chain orientation is augmented when solid specimens are heated to the LC state in an electric field and subsequently resolidified. This could be ascribed to the coupled effect of a preexisting molecular field and dielectric alignment of the mesogens.

The effects of AC fields on LCPs can be presented from the point of view of the dielectric permittivity. Thus, one can consider that its anisotropy can reveal information on the dipolar orientation. When the difference between the dielectric permittivity parallel and the one orthogonal to the mesogen is positive, these structural units are aligned in the same direction as the external field. Conversely, for negative values of dielectric anisotropy, the mesogens tend to orient perpendicular to the electric field (Brostow 1998). Dipoles aligned orthogonal to the mesogen director are more easily reoriented than dipoles along the director. The latter needs a rotation of the entire mesogen around the polymer main chain. Therefore, the transverse relaxation process takes place at higher frequencies than the longitudinal relaxation. Depending on the frequency values, the mesogens can exhibit a planar or homeotropic orientation. This aspect lies at the basis of "two frequency addressing" principle (Luckhurst and Veracini 1994, Seidel 2008).

The rotation of the mesogen around the main chain is known as the δ process. Depending on the polymer structure and conformation, this process can be narrow, as for one-comb-flexible polysiloxanes (Brostow 1998), or broad as observed for polyacrylates (Kresse et al. 1982). The δ process occurs faster in nematics than for smectic LCPs (Brostow 1998).

The response for the alignment of LCPs is affected by temperature and response time is proportional to reverse of the square of voltage applied to the material (Brostow 1998). As the glass transition temperature is higher, the response time is lower. At temperatures close to isotropization or in the biphasic region, one can achieve the fastest response times. This motivated scientists to orient LCPs during slow cooling through isotropic–nematic phase transition.

Processing of LCPs in the DC field is related to electrodynamic instabilities owing to the motion of ions. This limits the possibilities to accomplish a good alignment of LCPs, but through the so-called electrical cleaning, this issue is somehow solved (Platé 1993). The procedure is known to diminish the low-frequency loss and allows further alignment in an AC field. A drawback of the utilization of DC fields consists of injection of charges which deteriorate the samples and this is widely encountered for such alignments produced at temperatures close to glass transition temperature.

Optical properties of LCPs are strongly impacted by the presence of electrical fields since they induce birefringence and Kerr effect. Yan and coworkers (Yan et al. 2010) proposed a method to directly determine the electric-field-induced birefringence of a polymer-stabilized blue-phase LC (PS-BPLC) composite. The modification of refractive indices orthogonal to the electric field was assessed on Michelson interferometer, while the extraordinary refractive index of PS-BPLC was estimated from Senarmont method (Cloud 1998). The induced birefringence is linearly proportional to the square of the electric field intensity. As the electric field increases, the induced birefringence gradually saturates and deviates from Kerr effect. This phenomenon was called extended Kerr effect. The generated birefringence is almost three times the ordinary refractive index change. Lin and collaborators (Lin et al. 2011) have created another methodology for the estimation of induced birefringence and Kerr constant of PSBP-LC. The ordinary refractive and extraordinary refractive indices of PSBP-LC as a function of applied voltage can be determined experimentally from the phase shift of PSBP-LC and averaged refractive index at the voltage-off state. Consequently, the electric-field-induced birefringence and Kerr constant are calculated. This method is useful in the development of PSBP-LC-based photonic devices, such as displays, electro-optical switches, and tunable focusing lens arrays.

13.3.3 MAGNETIC FIELD

The effects created by magnetic fields on LCPs are similar to electric fields. Magnetic fields are more useful because there is no risk of partial discharges or breakdown. In the presence of a magnetic field, the macromolecules will tend to align along or against the field. The main condition is that the constituent molecules must have anisotropic magnetic susceptibility, but they should also be big enough to defeat the thermal disturbance.

The origin of the magnetic anisotropy is affected by the nature of the chemical bonds. For instance, the C–C bond is characterized by diamagnetic susceptibilities smaller in the direction of the bond than that normal to the bond (Brostow 1998). Thus, C–C bond mainly aligns orthogonal to the imposed field. Conversely, aromatic rings present a huge diamagnetic anisotropy because of the ring current generated inside it. As a result, this group is able to orient with the ring plane parallel to the applied field. These bonds are widely encountered in LCPs with longitudinal molecules, particularly those containing p-phenylene groups in the backbone. The effect is highlighted in LC phases when thermal randomizing effect is reduced. The polarization is generated by the magnetic field and is less found in the molecules as dipoles. In such situations, only quadrupolar orientation can occur (Brostow 1998).

The origin of the magnetic alignment of LCPs is a "domain" within which polymer chains can follow field direction and are cooperatively rotated by a magnetic torque (Coates 2000). The dimension and shape of the domain are not well-enough defined except for disclinations. Thus, one can visualize LCP systems as being composed of randomly oriented domains that can respond to the applied field independently. This is not widely valid because the size and the shape of the domains would range in agreement with the modification disclinations position during the magnetic alignment.

An approach that attempts to explain the LCP orientation kinetics under a magnetic field considers the fact that LCP fluid is multiphasic and comprises an interphase region where chain segments have greater mobility and less organization (Moore and Stupp 1987). For materials in nonequilibrium states, orientation dynamics of main-chain LCPs are equivalent with segments of a single chain going through various orientationally ordered domains.

The kinetics of the LCP chain alignment in the presence of magnetic fields was also reported by Anwer and Windle (1993). Their study was focused on poly(p-hydroxybenzoic acid-co-2,6-hydroxynaphthoic acid) of different molar masses, in a magnetic field applied at high temperatures corresponding to the nematic state. The order parameter of the LCP subjected to a magnetic field increased with time and finally reached a constant value. The retardation time was enhanced with lowering the magnetic field strength and increasing molar mass. The unoriented polymer presented slightly wandering director fields in three dimensions. The maximum orientation can be explained by assuming that all domains are entirely aligned and that the boundaries are not fully oriented. The remarked enhancement of order parameter with increasing field strength is thus due to lowering of the width of the domain boundaries with enhancement of field strength. The magnetic field can generate structural changes/transitions in LCP materials (Hocine et al. 2011, Parshin et al. 2012).

The effects of magnetic fields on flexural properties of a longitudinal LCP were reported (Brostow et al. 2002). A copolymer of poly(ethylene terephthalate + 0.6 mole fraction of *p*-hydroxybenzoic acid) in magnetic fields up to 1.8 T was investigated. The material was subjected to high temperatures until it reached the molten state and then it was cooled to the room temperature. During this procedure, the sample was placed under the magnetic field to maintain the orientation. Anisotropy increases with time and magnetic field strength, as a result of phase transitions. The manner in which orientation affects the mechanical properties of samples has been established by flexural tests. The strengthening effect is mainly influenced by the presence, size, and distribution of LC islands in the system. These LC islands behave similarly to dispersoids in other types of materials. Maxima of the strengthening effect with respect to the alignment are found and related to the size of LC-rich islands in the structure.

Utilization of a magnetic field of a specific profile influences the alignment profile of LCP chains in a film (Kimura 2003). A graded alignment can be achieved with an electromagnet with asymmetric pole pieces. Subjecting an LCP film (screwed between a set of heater plates) to a field of 2.4 T and also to cooling from 280°C, one can notice that the chains are aligned almost perpendicular to the film at the region near to the sample center. Above and below the center, the chains are tilted in regard with the direction perpendicular to the LCP surface. In the absence of external perturbation, the chains are aligned parallel to the film surface.

13.4   APPLICATIONS

LCPs applicability extends more and more as research into this field progresses. Practical uses of these materials range from the production of high-strength fibers to their implementation in optical devices.

The dynamic mechanical features of LCP fibers have been exploited in the industry of ballistics and protective clothing, where strong and lightweight materials are needed. Ordinary polymers do not exhibit the stiffness required to compete against some mechanically tough materials including metals. The properties of these materials are suited for the creation of products of various forms exposed to a wide range of threads of distinct geometric shapes being directed at the target. The latter is characterized by a variable dynamic impact profile from the point of view of speed and energy—aspects that are essential in the production of bulletproof vests. The heat resistance at elevated temperatures of aramid-derived LCPs is known to be very high for long periods of time. This makes such LCP materials good candidates for producing equipment for fire fighters.

Given their ability to interact with electromagnetic fields, LCPs are widely used in the display industry. At this point, scientists are focused on the enhancement of the time response to electric fields to achieve a faster alignment and implicitly to improve the ability of the display to change rapidly from one image to another. This problem is partially solved when LCPs are used since they are easier to manipulate comparatively with traditional LC. In the devices where response time is not so important, a twisted nematic LCP cell is utilized to produce energy-efficient displays. Prior to any external factor, the twisted LC phase reorients incident light which can pass through the second polarizer, resulting in a clear image to the observer. When LC layer is subjected to electric fields, the mesogen becomes aligned and is no longer twisting. In this stage, they are not reorienting incident radiation so that the flux of the polarized beam at the first polarizer is canceled by the second polarizer. Thus, the device appears dark to the observer. In this manner, an electric field can enable the creation of a pixel between dark and clear states on command. For rendering color to the image, color filters are used. All LCP-based optical devices are projected starting from this principle. Moreover, the mesogens can be locked in a specific configuration by selectively melting parts of the display into the LC phase, then subjecting the system to an electric field, and finally cooling down the polymer which hardens into a glass. This can improve the reliability of the device. Side-chain LCPs present remarkable properties for applications in optically nonlinear devices such as optical waveguides and electro-optic modulators in poled-slab waveguides based on polymers.

Other applications rely on the fact that the phase transitions caused by heating such materials can be exploited in thermography. For instance, cholesteric LCPs can change their color during heating in a certain temperature domain. The process is reversible and is used for thermal mapping of solid-state electronic devices or for monitoring human skin temperature variations.

13.5 GENERAL REMARKS

Main-chain LCPs can be oriented by applying external fields and then quenched to obtain a highly ordered and strong solid. As these technologies continue to progress, a wide variety of novel materials with strong and lightweight features will be found in various markets. This will significantly contribute to technical progress through the development of tunable notch filters, optical amplifiers, optically addressed spatial light modulators, and laser beam deflectors. The physical characteristics of ferroelectric chiral smectic C phases will open new perspectives in nonlinear optics.

ACKNOWLEDGMENT

This chapter was supported by grant of the Romanian National Authority for Scientific Research and Innovation, CNCS–UEFISCDI, project PN-II-RU-TE-2014-4-2976, no. 256/1.10.2015.

REFERENCES

Anwer, A. and Windle, A.H. 1993. Magnetic orientation and microstructure of main-chain thermotropic copolyesters. *Polymer* 34:3347–3357.
Barzic, A.I., Hulubei, C., Avadanei, M.I., Stoica, I., and Popovici, D. 2015. Polyimide precursor pattern induced by banded liquid crystal matrix: Effect of dianhydride moieties flexibility. *J. Mater. Sci.* 50:1358–1369.

Brostow, W. 1998. *Mechanical and Thermophysical Properties of Polymer Liquid Crystals.* Chapman & Hall, London.

Brostow, W., Jalewicz, M., Mehta, S., and Montemartini, P. 2002. Effects of magnetic fields on flexural properties of a longitudinal polymer liquid crystal. *Mater. Res. Innov.* 5:261–267.

Chen, R.H. 2011. *Liquid Crystal Displays: Fundamental Physics and Technology* Wiley, London.

Cloud, G.L. 1998. *Optical Methods of Engineering Analysis.* Cambridge University Press, New York.

Coates, D. 2000. Liquid crystal polymers: Synthesis, properties and applications. *Review Reports,* Report 118, 10(10), Shawbury, Rapra.

Collyer, A.A. 1992. *Liquid Crystal Polymers: From Structures to Applications.* Elsevier, New York.

Cosutchi, A.I., Hulubei, C., Stoica, I., and Ioan, S. 2010. Morphological and structural–rheological relationship in epiclon-based polyimide/hydroxypropylcellulose blend systems. *J. Polym. Res.* 17:541–550.

Cosutchi, A.I., Hulubei, C., Stoica, I., and Ioan, S. 2011. A new approach for patterning epiclon-based polyimide precursor films using a lyotropic liquid crystal template. *J. Polym. Res.* 18:2389–2402.

Dierking, I. 2014. A review of polymer-stabilized ferroelectric liquid crystals. *Materials* 7:3568–3587.

Ding, J., Feng, J., and Yang, Y. 1995. Sinusoidal supermolecular structure of band textures in a presheared hydroxypropyl cellulose film. *Polym. J.* 27:1132–1138.

Donald, A.M., Viney, C., and Windle, A.H. 1983. Banded structures in oriented thermotropic polymers. *Polymer* 24:155–159.

Donald, A.M. and Windle, A.H. 1992. *Liquid Crystalline Polymers.* Cambridge University Press, Cambridge.

Drzaic, P.S. 1995. *Liquid Crystal Dispersions.* World Scientific, Singapore.

Dutta, D., Fruitwala, H., Kohli, A., and Weiss, R.A. 1990. Polymer blends containing liquid crystals: A review. *Polym. Eng. Sci.* 30:1005–1018.

Fabre, P. and Veyssie, M. 1987. Shear viscosity experiment in side-chain nematic polymers. *Mol. Cryst. Liq. Cryst. Lett.* 4:99–105.

Graziano, D.J. and Mackley, M.R. 1984. Disclinations observed during the shear of MBBA. *Mol. Crys. Liq. Cryst.* 106:103–119.

Hocine, S., Brûlet, A., Jia, L., Yang, J., Di Cicco, A., Bouteiller, L., and Li, M.H. 2011. Structural changes in liquid crystal polymer vesicles induced by temperature variation and magnetic fields. *Soft Matter* 7:2613–2623.

Kimura, T. 2003. Study on the effect of magnetic fields on polymeric materials and its application. *Polym. J.* 35:823–843.

Kossyrev, P.A., Qi, J., Priezjev, N.V., and Crawford, G.P. 2002. Model of Freedericksz transition and hysteresis effect in polymer stabilized nematic liquid crystal configurations for display applications. *SID Symp. Dig. Tech. Pap.* 33:506–509.

Kresse, H., Kostromin, S., and Shibaev, V.P. 1982. Thermotropic liquid crystalline polymers, 10. Comparative dielectric investigations of smectic and nematic polymers. *Die Makromolekulare Chem.* 3:509–513.

Kulichikhin, V.G. 1988. Rheology, phase equilibria and processing of lyotropic liquid crystalline polymers. *Mol. Crys. Liq. Crys.* 169:51–81.

Lin, Y.H., Chen, H.S., Wu, C.H., and Hsu, H.K. 2011. Measuring electric-field-induced birefringence in polymer stabilized blue-phase liquid crystals based on phase shift measurements. *J. Appl. Phys.* 109:104503–104508.

Luckhurst, G.R. and Veracini, C.A. 1994. *The Molecular Dynamics of Liquid Crystals.* NATO ASI Series. Springer + Business Media, Dordrecht, Berlin.

Ma, R.Q. and Yang, D.K. 2000. Freedericksz transition in polymer-stabilized nematic liquid crystals. *Phys. Rev. E, Stat. Phys. Plasmas, Fluids, Relat. Interdiscip. Top.* 61:1567–1573.

Marrucci, G. and Greco, F. 1993. Flow behavior of liquid crystalline polymers. In *Advances in Chemical Physics*, eds. I. Prigogine and S.A. Rice, 331–404. Wiley, Hoboken.

Martin, P.G. and Stupp, S.I. 1987. Solidification of a main-chain liquid-crystal polymer: Effects of electric fields, surfaces and mesophase ageing. *Polymer* 28:897–906.

Moore, J.S. and Stupp, S.I. 1987. Orientation dynamics of main-chain liquid crystal polymers. 2. Structure and kinetics in a magnetic field. *Macromolecules* 20:282–293.

Navard, P. 1986. Formation of band textures in hydroxypropylcellulose liquid crystals. *J. Polym. Sci. Part B* 24:435–442.

Onogi, S. and Asada, T. 1990. Rheology and rheo-optics of polymer liquid crystals. In *Rheology*, vol. 1, eds. G. Astarita, G. Marrucci, and N. Luigi, 127–147. Springer, New York.

Parshin, A.M., Nazarova, V.G., Zyryanova, V.Y., and Shabanova, V.F. 2012. Magnetic-field-induced structural transition in polymer-dispersed liquid crystals. *Mol. Crys. Liq. Crys.* 557:50–59.

Platé, N.A. 1993. *Liquid-Crystal Polymers*. Springer, New York.

Scharf, T. 2007. *Polarized Light in Liquid Crystals and Polymers*. Wiley, New Jersey.

Scott, J.W., Jay, G.D., and Crawford, G.P. 2007. Liquid-crystal materials find a new order in biomedical applications. *Nat. Mater.* 6:929–938.

Seidel, A. 2008. *Characterization and Analysis of Polymer*. Wiley, Hoboken.

Shibaev, V.P. and Lam, L. 1994. *Liquid Crystalline and Mesomorphic Polymers*. Springer, New York.

Thakur, V.K. and Kessler, M.R. 2016. *Liquid Crystalline Polymers: Volume 1—Structure and Chemistry*. Springer, New York.

Werbowyj, R.S. and Gray, D.G. 1984. Optical properties of hydroxypropyl cellulose liquid crystals. I. Cholesteric pitch and polymer concentration. *Macromolecules* 17:1512–1520.

Yan, J., Jiao, M., Rao, L., and Wu, S.T. 2010. Direct measurement of electric-field-induced birefringence in a polymer-stabilized blue-phase liquid crystal composite. *Opt. Express* 18:11450–11455.

Index

Printed and bound by CPI Group (UK) Ltd, Croydon, CR0 4YY

01/11/2024

01782619-0006